AF410700

Recent Advances in the Understanding of Molecular Mechanisms of Resistance in Noctuid Pests

Recent Advances in the Understanding of Molecular Mechanisms of Resistance in Noctuid Pests

Editors

Gaelle Le Goff
Ralf Nauen

MDPI • Basel • Beijing • Wuhan • Barcelona • Belgrade • Manchester • Tokyo • Cluj • Tianjin

Editors
Gaelle Le Goff
INRAE, CNRS, Institut Sophia
Agrobiotech
Université Côte d'Azur
Sophia Antipolis
France

Ralf Nauen
Bayer AG, Crop Science Division
Molecular Entomology &
Toxicology
Monheim
Germany

Editorial Office
MDPI
St. Alban-Anlage 66
4052 Basel, Switzerland

This is a reprint of articles from the Special Issue published online in the open access journal *Insects* (ISSN 2075-4450) (available at: www.mdpi.com/journal/insects/special_issues/molecular_resistance_Noctuid).

For citation purposes, cite each article independently as indicated on the article page online and as indicated below:

LastName, A.A.; LastName, B.B.; LastName, C.C. Article Title. *Journal Name* **Year**, *Volume Number*, Page Range.

ISBN 978-3-0365-1758-2 (Hbk)
ISBN 978-3-0365-1757-5 (PDF)

Contents

About the Editors

Gaelle Le Goff

Gaëlle Le Goff is a senior scientist at the Institut Sophia Agrobiotech, French National Institute for Agriculture, Food and Environment (INRAE). She has a PhD in Molecular Biology and Pharmacology from the University of Nice Sophia Antipolis, France. She did a post-doc at the University of Bath in UK with a European Marie Curie grant and was then recruited at INRA, where she still carries out her research in functional genomics in insects. Her current projects focus on i) insect adaptation to their chemical environment (adaptation to plant secondary metabolites/insecticide resistance), ii) regulation of detoxification genes expression, and iii) annotation of insect genomes (detoxification genes).

Ralf Nauen

Ralf Nauen is an insect toxicologist working on functional (toxico)genomics, molecular entomology, fundamental and applied aspects of insecticide/acaricide mode of action, selectivity and detoxification, as well as biochemical and molecular mechanisms of insecticide resistance and its management. He received his PhD from the University of Portsmouth (UK) and is a Bayer Distinguished Science Fellow. In 2013 he was awarded to become Fellow of the Entomological Society of America and in 2014 he received the prestigious American Chemical Society International Award for Research in Agrochemicals, in recognition of his outstanding and influential research into insecticide and acaricide modes of action and resistance. He authored more than 230 scientific papers/book chapters with more than 22,000 citations. He received the Highly Cited Researcher award by the Web of Science 2018-2020. He is appointed Visiting Professor by the Chinese Academy of Agricultural Sciences (Beijing, China) and lecturer at the University of Bonn (Germany).

Editorial

Recent Advances in the Understanding of Molecular Mechanisms of Resistance in Noctuid Pests

Gaëlle Le Goff [1],* and **Ralf Nauen** [2],*

1   Université Côte d'Azur, INRAE, CNRS, ISA, F-06903 Sophia Antipolis, France
2   Bayer AG, Crop Science Division, R&D Pest Control, Alfred Nobel Str. 50, 40789 Monheim, Germany
*   Correspondence: gaelle.le-goff@inrae.fr (G.L.G.); ralf.nauen@bayer.com (R.N.); Tel.: +33-4-9238-6578 (G.L.G.); +49-21-7338-4441 (R.N.)

Citation: Le Goff, G.; Nauen, R. Recent Advances in the Understanding of Molecular Mechanisms of Resistance in Noctuid Pests. *Insects* **2021**, *12*, 674. https://doi.org/10.3390/insects12080674

Received: 16 July 2021
Accepted: 24 July 2021
Published: 27 July 2021

Publisher's Note: MDPI stays neutral with regard to jurisdictional claims in published maps and institutional affiliations.

Noctuid moths are among the most devastating crop pests on the planet. Many of them are polyphagous pests capable of feeding on numerous host plants. Their larval stages cause significant damage to plants of agronomic importance, including corn, cotton, soybean, and rice, which result in severe yield losses if not kept under economic damage thresholds. Control of these pest insects is mainly based on the use of chemical insecticides and/or the cultivation of transgenic plants expressing *Bacillus thuringiensis* (Bt) pore-forming insecticidal proteins (Cry toxins). However, the efficacy of these control measures is often jeopardized by the development of resistance due to frequent applications of insecticides and long-term exposure to a relatively small set of Cry toxins expressed in transgenic crops. Understanding the molecular mechanisms of resistance that these pest insects have developed is essential for the implementation of sustainable control methods and resistance management strategies. This special issue is dedicated to this topic and includes 10 papers presenting either original results or reviews.

Several species of the noctuid moth family are invasive species that have recently spread to various parts of the world. The most problematic is probably *Spodoptera frugiperda* (Smith, 1797), which originated from the American continent and recently invaded the Western hemisphere. It was first detected in Africa in 2016 and later in Asia (2018) and Australia (2020). It is approaching Europe and was detected in the Canary Islands (Spain) in July 2020. The Food and Agriculture Organization (FAO) has this year proclaimed it as "one of the most destructive pests jeopardizing food security across vast regions of the globe". The importance of this species, also known as the fall armyworm (FAW), is reflected in the Special Issue, with half of the publications dedicated to it. Some of the papers focus on the diagnosis to differentiate the two sympatric host plant species of FAW, maize and rice [1], or on the most suitable rearing conditions for this non-model species [2], while others tackle the development of molecular tools for the detection of mutations in insecticide targets conferring resistance [3]. Other studies have monitored the presence of known mutations in the targets of major insecticide classes by comparing native and invasive populations [1,4]. Marked differences exist between FAW populations, which can provide guidance for appropriate resistance management decisions. Indeed, these studies show the presence of mutations in acetylcholinesterase, the target of carbamates and organophosphates, in both native and invasive populations, whereas only native populations show a deletion in ABCC2, the target of some Bt proteins [1,4]. In this case, diamide insecticides could be useful in the fight against FAW since none of the known mutations in the target of diamides, the ryanodine receptor, have been detected. However, it should not be forgotten that resistance can also be acquired through increased capacity to metabolize insecticides and involves detoxification enzymes, such as cytochromes P450 monooxygenases, carboxylesterases, glutathione S-transferases, or transporters like ATP binding cassette (ABC) transporters. Some of the roles of these detoxification genes are highlighted in a review on species in the genus *Spodoptera* [5]. Nevertheless, the fight against insect pests must also include research and discovery of new molecules with different modes of action to counteract resistance mechanisms. Such research should also take into account a societal demand for the development of more environmentally friendly molecules [6].

The other noctuid moth species covered in this Special Issue are also important pests. Klai et al. proposed an analysis of the presence of transposable elements in the genome of *Helicoverpa armigera* and focused on their presence in detoxification genes as they may cause changes in the expression level and confer resistance [7]. Other authors have sought to develop molecular tools either to differentiate between two closely related species, such as *Mythimna loreyi* and *Mythimna separata*, with the LAMP (loop-mediated isothermal amplification) assay [8], or by designing primers to monitor the expression of Bt toxin target genes and reference genes in soybean looper, *Chrysodeixis includens* [9].

ABC transporters are important Bt protein receptors and there is an increasing body of evidence that mutations in ABC transporters (e.g., ABCC2) confer high levels of resistance to Cry toxins as reviewed in [10]. In his review, Heckel [10] highlights a number of research questions that remain open, such as how potential Cry toxin receptor proteins and Cry toxins interact to form pores. However, our knowledge and understanding regarding the function and regulation of genes conferring resistance in pests is constantly increasing and particularly facilitated by the exploitation of new technologies, such as genome editing by CRISPR/Cas9.

Finally, we would like to thank all authors for their commitment to contribute their articles and the many reviewers for their valuable comments and suggestions helping all of us to shape this interesting Special Issue, which hopefully attracts a broad readership of *Insects*.

Institutional Review Board Statement: Not applicable.

Informed Consent Statement: Not applicable.

Data Availability Statement: Not applicable.

Conflicts of Interest: The authors declare no conflict of interest.

References

1. Boaventura, D.; Martin, M.; Pozzebon, A.; Mota-Sanchez, D.; Nauen, R. Monitoring of Target-Site Mutations Conferring Insecticide Resistance in *Spodoptera frugiperda*. *Insects* **2020**, *11*, 545. [CrossRef] [PubMed]
2. Gergs, A.; Baden, C. A Dynamic Energy Budget Approach for the Prediction of Development Times and Variability in *Spodoptera frugiperda* Rearing. *Insects* **2021**, *12*, 300. [CrossRef] [PubMed]
3. Schlum, K.; Lamour, K.; Tandy, P.; Emrich, S.; de Bortoli, C.; Rao, T.; Dillon, D.V.; Linares-Ramirez, A.; Jurat-Fuentes, J. Genetic Screening to Identify Candidate Resistance Alleles to Cry1F Corn in Fall Armyworm Using Targeted Sequencing. *Insects* **2021**, *12*, 618. [CrossRef]
4. Yainna, S.; Nègre, N.; Silvie, P.; Brévault, T.; Tay, W.; Gordon, K.; Dalençon, E.; Walsh, T.; Nam, K. Geographic Monitoring of Insecticide Resistance Mutations in Native and Invasive Populations of the Fall Armyworm. *Insects* **2021**, *12*, 468. [CrossRef] [PubMed]
5. Hilliou, F.; Chertemps, T.; Maïbèche, M.; Le Goff, G. Resistance in the Genus *Spodoptera*: Key Insect Detoxification Genes. *Insects* **2021**, *12*, 544. [CrossRef] [PubMed]
6. Ruttanaphan, T.; De Sousa, G.; Pengsook, A.; Pluempanupat, W.; Huditz, H.-I.; Bullangpoti, V.; Le Goff, G. A Novel Insecticidal Molecule Extracted from *Alpinia galanga* with Potential to Control the Pest Insect *Spodoptera frugiperda*. *Insects* **2020**, *11*, 686. [CrossRef] [PubMed]
7. Klai, K.; Chénais, B.; Zidi, M.; Djebbi, S.; Caruso, A.; Denis, F.; Confais, J.; Badawi, M.; Casse, N.; Khemakhem, M.M. Screening of *Helicoverpa armigera* Mobilome Revealed Transposable Element Insertions in Insecticide Resistance Genes. *Insects* **2020**, *11*, 879. [CrossRef] [PubMed]
8. Nam, H.Y.; Kwon, M.; Kim, H.J.; Kim, J. Development of a Species Diagnostic Molecular Tool for an Invasive Pest, *Mythimna loreyi*, Using LAMP. *Insects* **2020**, *11*, 817. [CrossRef] [PubMed]
9. Martin, M.; Boaventura, D.; Nauen, R. Evaluation of Reference Genes and Expression Level of Genes Potentially Involved in the Mode of Action of Cry1Ac and Cry1F in a Susceptible Reference Strain of *Chrysodeixis includens*. *Insects* **2021**, *12*, 598. [CrossRef] [PubMed]
10. Heckel, D. The Essential and Enigmatic Role of ABC Transporters in Bt Resistance of Noctuids and Other Insect Pests of Agriculture. *Insects* **2021**, *12*, 389. [CrossRef] [PubMed]

insects

MDPI

Review

The Essential and Enigmatic Role of ABC Transporters in Bt Resistance of Noctuids and Other Insect Pests of Agriculture

David G. Heckel

Max Planck Institute for Chemical Ecology, Hans-Knoell-Str. 8, D-07745 Jena, Germany; heckel@ice.mpg.de; Tel.: +49-3641-57-1599

Simple Summary: The insect family, Noctuidae, contains some of the most damaging pests of agriculture, including bollworms, budworms, and armyworms. Transgenic cotton and maize expressing Cry-type insecticidal proteins from *Bacillus thuringiensis* (Bt) are protected from such pests and greatly reduce the need for chemical insecticides. However, evolution of Bt resistance in the insects threatens the sustainability of this environmentally beneficial pest control strategy. Understanding the interaction between Bt toxins and their targets in the insect midgut is necessary to evaluate the risk of resistance evolution. ABC transporters, which in eukaryotes typically expel small molecules from cells, have recently been proposed as a target for the pore-forming Cry toxins. Here we review the literature surrounding this hypothesis in noctuids and other insects. Appreciation of the critical role of ABC transporters will be useful in discovering counterstrategies to resistance, which is already evolving in some field populations of noctuids and other insects.

Abstract: In the last ten years, ABC transporters have emerged as unexpected yet significant contributors to pest resistance to insecticidal pore-forming proteins from *Bacillus thuringiensis* (Bt). Evidence includes the presence of mutations in resistant insects, heterologous expression to probe interactions with the three-domain Cry toxins, and CRISPR/Cas9 knockouts. Yet the mechanisms by which ABC transporters facilitate pore formation remain obscure. The three major classes of Cry toxins used in agriculture have been found to target the three major classes of ABC transporters, which requires a mechanistic explanation. Many other families of bacterial pore-forming toxins exhibit conformational changes in their mode of action, which are not yet described for the Cry toxins. Three-dimensional structures of the relevant ABC transporters, the multimeric pore in the membrane, and other proteins that assist in the process are required to test the hypothesis that the ATP-switch mechanism provides a motive force that drives Cry toxins into the membrane. Knowledge of the mechanism of pore insertion will be required to combat the resistance that is now evolving in field populations of insects, including noctuids.

Keywords: *Bacillus thuringiensis*; Cry protein; ATP-Binding Cassette; ABC Transporter; ATP switch model; pore-forming toxin; resistance; genetics; Noctuidae; *Helicoverpa*; *Spodoptera*; *Heliothis*; *Chloridea*; *Trichoplusia*

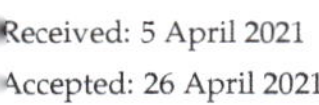

Citation: Heckel, D.G. The Essential and Enigmatic Role of ABC Transporters in Bt Resistance of Noctuids and Other Insect Pests of Agriculture. *Insects* **2021**, *12*, 389. https://doi.org/10.3390/insects12050389

Academic Editors: Gaelle Le Goff and Ralf Nauen

Received: 5 April 2021
Accepted: 26 April 2021
Published: 28 April 2021

Publisher's Note: MDPI stays neutral with regard to jurisdictional claims in published maps and institutional affiliations.

1. Introduction

ABC proteins are a huge and ancient superfamily of proteins that are defined by the presence of a domain called the ATP-binding cassette [1]. Binding to ATP causes this domain to change its shape, and this conformational change has been harnessed for many purposes. The ABCE1 protein is required for termination of protein translation, where its lever-like action separates the large and small ribosomal subunits [2]. The Rad50 protein is invoked when chromosomes are damaged, by clamping onto DNA like tweezers to bring the strands together for repair [3]. The ABC transporters possess membrane-spanning helices that change orientation to squeeze small molecules across membranes when the ABC domain binds to ATP [4]. The most recently discovered property of ABC proteins is their role in the mode of action of pore-forming three-domain Cry toxins from *Bacillu thuringiensis* (Bt), the subject of this review.

The Cry toxins are important in agriculture because cotton, maize, and other crop have been transformed with genes from Bt to enable the plants to manufacture enougl toxins to kill certain pest caterpillars that feed on them. See [5] for a comprehensive list o Bt toxins. When ingested along with plant tissue, the toxins form pores in the larval midgu epithelium, lysing the cells and eventually killing the insect. In 2019, 108.7 million hectare of transgenic Bt crops were grown worldwide [6], greatly reducing the use of chemica insecticides. The sustainability of this strategy is threatened by the evolution of resistance which is already appearing in field populations of some pest insects (see [7] for a usefu overview). Studies of Bt-resistant insects provided the first evidence for ABC transporter in the Cry toxin mode of action.

Cry1-type toxins are particularly potent against polyphagous noctuid pests, primaril the heliothines and the armyworms. The cotton bollworm *Helicoverpa armigera* (Hübner was formerly distributed in Asia, Africa, and Australia, but has recently invaded Soutl America. Its resistance to chemical insecticides and its ability to hybridize with the nativ corn earworm *Helicoverpa zea* (Boddie) has promoted its spread. *H. zea* populations in Nortl America are evolving resistance to Cry1Ab-expressing maize. *Chloridea* (formerly *Heliothis virescens* (Fabricius), the tobacco budworm [8], is a pest of cotton and other crops that ha been well-controlled by Bt-cotton up to now. We will use the genus name *Heliothis* in thi review for continuity with the literature cited. The fall armyworm *Spodoptera frugiperd* (J. E. Smith) has recently invaded Africa, India, Asia, and Australia from its home in the New World to become a worldwide pest of maize. Its congeners, the beet armyworn *Spodoptera exigua* (Hübner) and cotton leafworm *Spodoptera litura* (Fabricius), are als polyphagous pests. The cabbage looper *Trichoplusia ni* (Hübner) has evolved resistanc to Dipel in greenhouses. Cry1 toxins are also effective against non-noctuid Lepidopter such as the diamondback moth *Plutella xylostella* (L), a worldwide pest of crucifer crop and the first insect to evolve Bt resistance in open field populations, and the domesticatec silkworm *Bombyx mori* (L), a model insect for which powerful genetic tools have beer developed. Cry1 resistance has appeared in some populations of the European corn bore *Ostrinia nubilalis* (Hübner) and the Asian corn borer *Ostrinia furnacalis* (Guenée). Cry: toxins are active against many of the same Lepidoptera. The pink bollworm *Pectinophor gossypiella* has evolved resistance to Cry2-expressing Bt-transgenic cotton. The Cry3 toxin are potent against Coleoptera, such as *Diabrotica virgifera* (LeConte), the corn rootworn ABC transporters of these non-noctuids will also be discussed as they illustrate the commo features of the mode of action of the three-domain Cry toxins. Cell lines from noctuids hav also been important in Bt research, including Sl-HP from *S. litura*, Sf9 from *S. frugiperd* Hi5 from *T. ni*, and QB-Ha-E5 from *H. armigera*.

Unfortunately, there is no standard nomenclature yet for insect ABC transporter which is becoming a problem for comparative studies as more and more are implicatec as Bt targets. Gene and protein names provided by automatic annotation at NCBI do nc uniquely identify them or reveal their relationship to the ABC transporters described her The most reliable point of reference is the genome of *B. mori*, the first to be sequencec among Lepidoptera. Figure 1 shows the genomic context of the *B. mori* orthologs of the proteins discussed in this review. Figure 2 shows their predicted membrane topology.

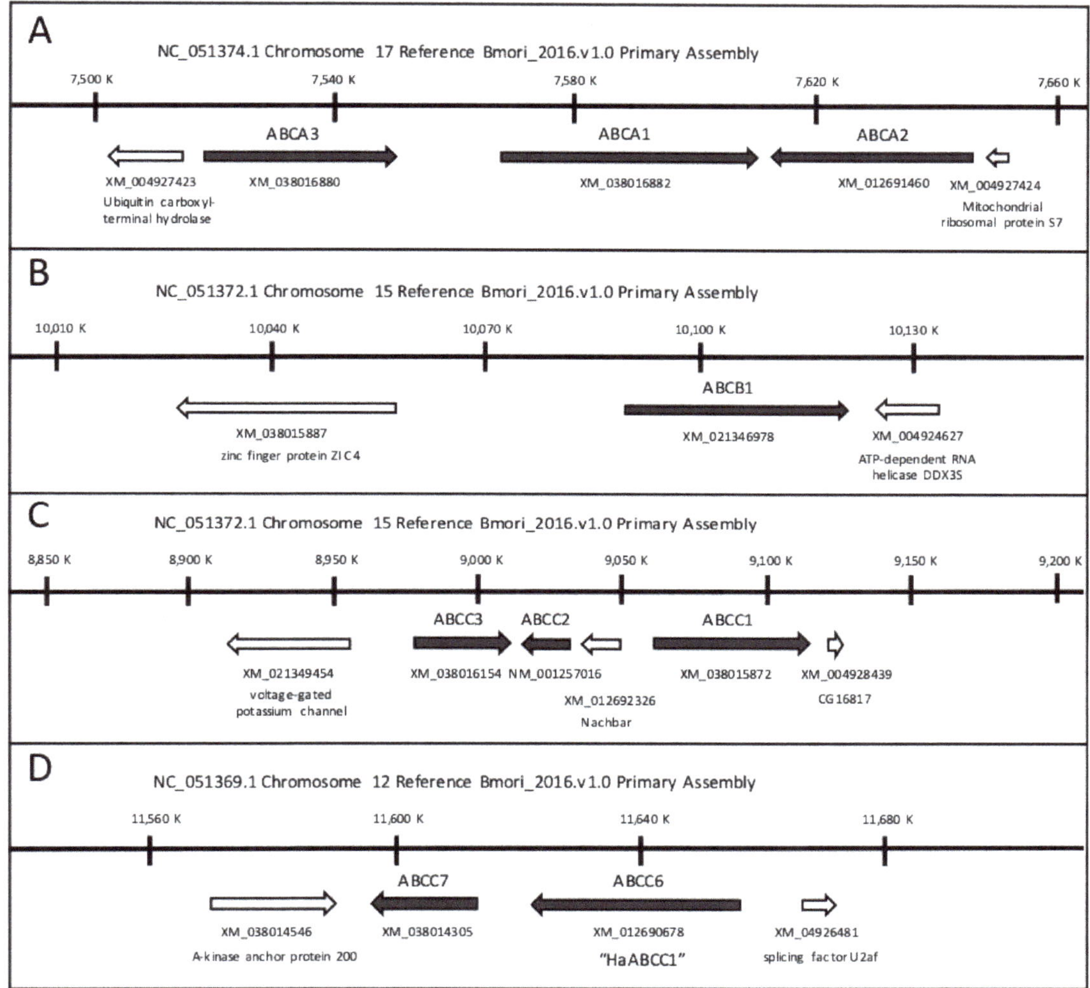

Figure 1. Genomic location of *Bombyx mori* (**A**–**C**) transporters orthologous to those interacting with three-domain Cry proteins described in the text. Coordinates are based on the chromosome view on NCBI (https://www.ncbi.nlm.nih.gov/, accessed on 2 April 2021) and are the same as shown in the new version of Kaikobase [9]. (**D**) Names for ABCC6 and ABCC7 are defined here for the first time, to avoid confusion with the previously named ABCC1.

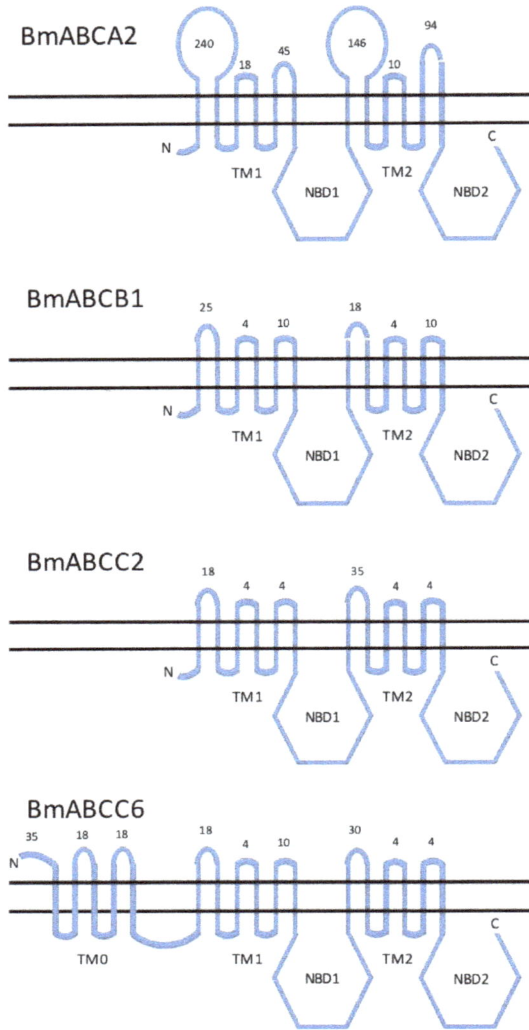

Figure 2. Phobius [10,11] predictions of membrane topology in the lipid bilayer of four ABC transporters from *Bombyx mori* orthologous to those interacting with three-domain Cry proteins described in the text (not to scale). The external lumenal surface is on top. Numbers of residues in each predicted external loop are shown. Transmembrane domains: TM0, TM1, and TM2. Nucleotide-binding domains: NBD1, NBD2. The N-terminus and C-terminus of each polypeptide is indicated.

2. Initial Discoveries

ABC transporters were first discovered as targets of Bt toxins by positional cloning in strains of resistant insects. Despite decades of research on Bt resistance, there was no previous biochemical or physiological evidence that ABC transporters could be involved. The discovery of their role was a surprise, and only the independent efforts by three groups that converged on the same protein has convinced the research community that understanding their role in resistance is important, as described in several recent reviews [5,12–23]. Solving this problem holds the key to developing strategies to combat the increasing problem of Bt resistance.

A mutation in the ABCC2 protein was identified in a Cry1Ac-resistant strain of *H. virescens*, by positional cloning using markers from the early versions of the *B. mori* genome sequence, well before a genome sequence for *H. virescens* was available [24]. Evidence that this mutation was important for resistance came from mapping, binding studies, and an allele frequency change correlated with the increase of resistance over time [24]. Like most ABC mutations subsequently found in other species, and like the cadherin mutation previously found in *H. virescens* [25] it introduced a frameshift and prevented expression of the full-length protein in the membrane. This contrasts with many cases of chemical insecticide resistance, where deletion of the target would be lethal.

Even stronger evidence came from analysis of a more subtle mutation in the ABCC2 protein in *B. mori* [26]. Positional cloning using the genome sequence converged on ABCC2, but no incapacitating mutation could be found. Instead, the protein in the Cry1Ab-resistant strain had several amino acid substitutions and an insertion of a tyrosine in an extracellular loop. Germline transformation was used to prove that one copy of the susceptible allele inserted in the resistant genetic background could confer Cry1Ab-susceptibility on the resistant strain [26], and subsequent experiments [27] proved that the inserted tyrosine was necessary and sufficient for resistance.

A similar approach using bulked segregant analysis based on cDNA markers identified ABCC2 and its neighboring gene ABCC3 as contributing to Cry1Ac and Cry1Ca resistance in *S. exigua* [28]. In contrast to the previous mutations, ABCC2 carried a lesion in an intracellular nucleotide-binding domain (NBD). Suppression of ABCC2 and ABCC3 by RNA inhibition (RNAi) increased the tolerance of susceptible larvae to Cry1Ac and Cry1Ca. This independent and unbiased positional cloning approach extended the phenomenon to a third genus, and to a third type of mutation in ABCC2, proving that ABC transporters could no longer be ignored in the mode of action of Cry toxins.

3. Search for ABC Mutants in Resistant Strains from Field and Laboratory

These early findings motivated the search for the involvement of ABC transporters in other Bt-resistant strains of noctuids and other Lepidoptera. Comparative linkage mapping with markers, previously shown to be linked to ABCC2 in *H. virescens*, was used to identify a mutation that eliminated the last transmembrane domain of ABCC2 in Cry1Ac-resistant *P. xylostella* from Hawaii [29]. The same study localized Cry1Ac resistance in *T. ni* from British Columbia to a region containing ABCC2, although a specific mutation was not identified [29]; a different resistance mechanism, altered aminopeptidase expression, was also identified in the same strain of *T. ni* [30]. Mis-spliced transcripts of ABCC2 generating a truncated protein were found in a Cry1Ac-resistant strain of *H. armigera* from China [31]. Another comparative linkage mapping approach identified a genomic region containing ABCC2 in a laboratory-selected Cry1F-resistant strain of *O. nubilalis* from collections in the Corn Belt of the USA [32], although involvement of ABCC2 has not yet been confirmed in field-evolved Cry1F resistance in *O. nubilalis* from Nova Scotia [33]. In Puerto Rico, rapid appearance of Cry1F resistance in *S. frugiperda* stimulated withdrawal of the transgenic maize variety from the market, and was found to be associated with mutations in ABCC2 [34]. Additional mutations in *S. frugiperda* ABCC2 were associated with Cry1Fa and Cry1A.105 resistance in Puerto Rico [35] and Brazil [36]. Screening for some of these was included in surveys using DNA diagnostics for resistance to chemical insecticides as well as Bt [37,38].

The only member of the ABCB family to be investigated in Lepidoptera as a target of Cry toxins initially came to attention because it was down-regulated in a Cry1Ac-resistant strain of *P. xylostella*. PxABCB1 expression was found to be lower in other resistant strains, was further reduced by additional Cry1Ac selection, and was suppressed by RNAi in susceptible strains, which increased their tolerance to Cry1Ac [39].

4. Functional Studies

Heterologous expression of ABC transporters in insect cell lines has been extensively used to probe their function. The crucial role of the tyrosine insertion in loop 2 of *B. mori* ABCC2 in conferring Cry1Ab resistance was convincingly shown by expression in Sf9 cells [27]. The same study demonstrated the synergy of the cadherin BtR175 and ABCC2 for the first time; co-expression of both made the cells much more susceptible to Cry1Ab than either one alone. These results were recapitulated using proteins from *H. virescens* in Sf9 cells [40], with the added information that synergy was observed only when both proteins were expressed in the same cell; i.e., not from a mixture of cells expressing one or the other, as might be expected from the sequential binding hypothesis [41]. The mechanism of synergy was further probed by comparing the ability of the cadherin from *H. armigera* or *S. litura* to synergize Cry1Ac action on *H. armigera* ABCC2 expressed in Hi5 cells [42]. Although the *S. litura* cadherin was an ineffective synergist, when cadherin repeat 11 from *H. armigera* was swapped in, synergistic activity increased. The authors hypothesized that specific binding sites on the cadherin localized the toxin to a good position for interaction with ABCC2 in a species-specific manner [42]. A similar species-specific synergism with ABCC2 from *S. exigua* and the cadherin from *S. exigua* (but not *H. armigera*) was observed with the Cry1C toxin in Sf9 cells [43]. Domain-swapping between Cry1C and Cry1Ac was used to infer that domains II and III of Cry1Ac have different binding sites on ABCC2 of *S. exigua* [44]. ABCC2 from *S. frugiperda* expressed in Hi5 cells conferred sensitivity to Cry1Ab and Cry1Fa, while the cadherin did not, but synergism was not investigated in this study [45]. ABCC3 from *S. frugiperda* was also found to confer sensitivity to Cry1Ab and Cry1Fa under similar conditions [46].

A wide-ranging study explored the specificity of the toxin-target interaction by expressing ABC transporters from Lepidoptera, Coleoptera, Diptera, and humans in Sf9 cells and testing them with lepidopteran- or coleopteran-active toxins [47]. ABCC2 or ABCC3 from *B. mori* conferred sensitivity to Cry1Aa, but not Cry1Ca or Cry1Da. The latter two must have different, unknown targets because they are active on caterpillars of some lepidopteran species. Human and dipteran ABC transporters tested did not respond to lepidopteran- or coleopteran-active toxins.

D. melanogaster is not normally susceptible to Cry1Ac, but when ABCC2 was expressed in the midgut of transgenic larvae, Cry1Ac in the artificial diet killed them [48]. Moreover, the genome of *D. melanogaster* lacks the ortholog of the 12-cadherin domain protein found in all Lepidoptera, so the killing mechanism did not rely on the same type of synergism from the cadherin. However, synergism could be observed when the transgenic larvae were fed peptide fragments from lepidopteran cadherins along with Cry1Ac [48].

The most sensitive measurements of the interaction between Cry toxins and their receptors have been made using heterologous expression in *Xenopus* oocytes [49]. Messenger RNA experimentally injected into these huge cells is translated and the protein (e.g., ABCC2 or cadherin) are incorporated into the egg membrane. This technique is often used to investigate the properties of ion channels using the voltage-clamp technique. The current through the channel is measured as a function of the experimentally-fixed voltage gradient across the membrane and the resulting graph characterizes the electrophysiological properties of the channel. In this case, the channel is the Cry toxin pore inserted into the membrane, which allows inward cation flux. The dynamics of current flow depend on the details of pore insertion and structure. Using this sensitive technique, it was shown that expression of the cadherin alone produced almost no current, expression of ABCC2 allowed abundant current, and expression of both produced even more current—the most convincing demonstration of synergism to date.

The mechanism of synergism is still obscure, but a number of hypotheses can be envisaged, which are not mutually exclusive (Figure 3). These can be classified into *trans*-acting mechanisms where synergism can occur when the cadherin and ABC transporter may be separated from each other, and *cis*-acting mechanisms where synergism requires a close physical interaction. According to the sequential binding hypothesis [41], toxin

monomers bind to the cadherin and are further processed by cleavage of the N-terminal
α1-helix, whereupon they form oligomeric pre-pores in solution (Figure 3A). However,
toxin monomer binding to the cadherin is not an absolute requirement for toxicity; cadherin
knockouts can still be killed by higher toxin amounts [24,50–52] and Cry1Ac is still lethal
to *T. ni* dispite not being able to bind to the *T. ni* cadherin [53]. Synergism is due to
presence of the cadherin, which speeds up a process that happens at a slower rate in its
absence. Here the pre-pore structure can diffuse away from the cadherin to interact with
a remote ABC transporter, so this mechanism is classified as *trans*-acting. If, however
the cadherin traps the toxin and brings it to the ABC transporter, this would be a *cis*-
acting mechanism (Figure 3B). This could be synergistic if there are many more cadherin
molecules in the membrane than ABCC2 molecules. Another previously suggested *cis*-
acting mechanism [40] is shown in Figure 3C, where the cadherin pulls the pre-pore away
from the ABC transporter, enabling it to insert into the membrane and freeing up the ABC
transporter for the next pre-pore.

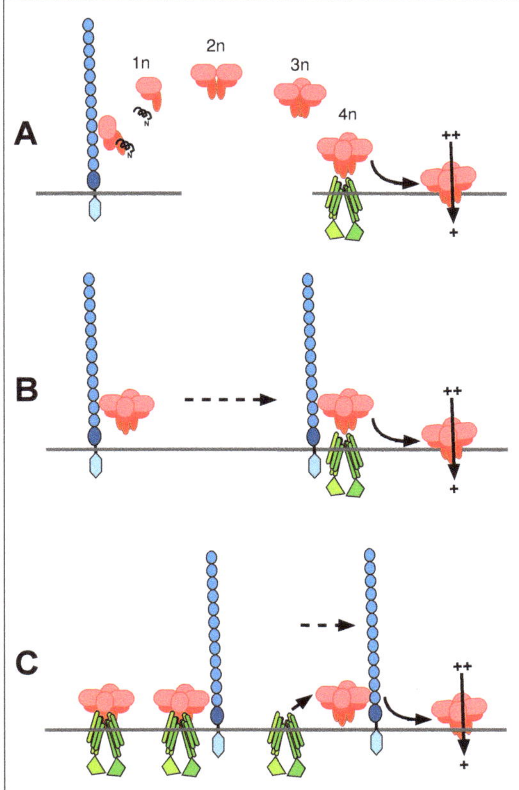

Figure 3. Hypothetical models of the mechanism of synergy between the 12-cadherin domain
protein and an ABC transporter. (**A**) *Trans*-acting synergy, due to acceleration of oligomer formation
following monomer binding to the cadherin. (**B**) *Cis*-acting synergy, due to the cadherin trapping the
pre-pore and moving it to the ABC transporter. (**C**) *Cis*-acting synergy, due to the cadherin pulling
the pre-pore away from the ABC transporter, freeing it to interact with another toxin pre-pore. Dotted
lines represent movement of the cadherin within the membrane.

5. Extracellular Loops

Most of the ABC protein is out of reach of Bt toxins approaching the cell, since the nucleotide binding domains are entirely cytoplasmic, and most of the transmembrane domains are buried within the lipid bilayer (Figure 2). Extracellular loops connecting adjacent transmembrane helices are very short in ABCC proteins, but larger in ABCB and especially ABCA proteins. Detailed analyses of the interaction between extracellular loops of *B. mori* ABC transporters and various domains of Cry1A toxins have been carried out by the group of Ryoichi Sato in Tokyo. Following the discovery that insertion of a single tyrosine in Loop 2 of ABCC2 conferred resistance to Cry1Ab in larvae, swapping other amino acids for the inserted tyrosine blocked pore formation in cells expressing the transporter, while amino acid substitutions at other positions in the non-inserted loop did not [54]. Thus, the size of the loop, rather than its amino acid composition, was the more important determinant of sensitivity. Domain-swapping within the toxin implicated Domain II as most important in this interaction. Subsequent mutagenesis of Domain II of Cry1Aa revealed a region that bound both to ABCC2 and another important receptor, the cadherin BtR175 [55]. While ABCC3 had much lower binding affinities to Cry1Aa and Cry1Ab than ABCC2, binding was increased in constructs containing partial loops from ABCC2 [56]. Another group pointed out an amino acid difference within Loop 1 of ABCC2 of *S. frugiperda* and *S. litura* that correlated with binding affinity to Cry1Ac, and they also replicated the binding difference by creating two versions of the *H. armigera* ABCC2, one with each amino acid [57].

6. Regulatory Changes

Regulatory changes involving ABC transporters were also found to confer resistance. In a Cry1Ac-resistant strain of *P. xylostella*, resistance mapped to the vicinity of ABCC2 but no disruptive mutations in the gene could be found [58]. Instead, the expression level of ABCC2 and ABCC3 was found to be controlled by the mitogen-activated protein kinase (MAPK) signaling pathway, with the MAP4K4 gene located close by on the same genomic scaffold, accounting for the mapping results. Constitutive expression of MAP4K4 in the resistant strain suppressed ABCC2, ABCC3, and alkaline phosphatase, another Cry1Ac-binding protein. RNA interference (RNAi) suppression of MAP4K4 transiently restored susceptibility by upregulating ABCC2 and ABCC3. Thus, resistance in this case was due to higher expression of a negative regulator of ABCC2 transcription.

The Forkhead Box Protein A (FOXA) transcription factor was found to stimulate transcription of ABCC2 and ABCC3 in *H. armigera*, as predicted from FOXA binding sites in the promoters [59]. RNAi silencing of FOXA downregulated ABCC2 and ABCC3 and increased the tolerance of susceptible larvae to Cry1Ac. Parallel results were obtained by expression in Sl-HP cells in the same study. A different study screened several members of the GATA transcription factor family from *H. armigera* and found that GATAe caused Sf9, QB-Ha-E5, and Hi5 cell lines to increase their expression of ABCC2, conferring greater Cry1Ac sensitivity [60]. If either mechanism were to be found in resistant strains from the field, resistance would be due to lower expression of a positive regulator of ABCC2 transcription.

Comparison of the ABCC2 coding sequence across many species of Lepidoptera identified a conserved region targeted by the microRNA miR-998-3p [61]. MicroRNAs bind to messenger RNAs in a sequence-specific fashion and target them for destruction or inhibit translation. Injection of an agomir (a chemically modified RNA that mimics the effect of the miRNA) into susceptible larvae of *P. xylostella*, *S. exigua*, or *H. armigera* increased their tolerance of Cry1Ac and decreased the abundance of ABCC2 mRNA. Injection of an antagomir (a single-stranded molecule designed to block the effect of the miRNA) into larvae of Cry1Ac-resistant *P. xylostella* reduced their tolerance of Cry1Ac and increased their ABCC2 mRNA levels.

7. Cell Lines

The influential colloid-osmotic lysis theory to explain how Cry toxins kill cells was developed using different cell lines that naturally differed in their susceptibilities to two different toxins [62]. It would be interesting to determine which ABC transporters are naturally expressed by those cells. Sl-HP cells from *S. litura* are susceptible to activated Cry1Ac even though *S. litura* larvae are not. ABCC3 was found to be expressed in this cell line, and RNA inhibition of ABCC3 decreased Sl-HP sensitivity to Cry1Ac [63]. In another study, comparison of Cry1Ac sensitivities of cell lines from different tissues produced the order midgut > fat body > ovary as expected, but unexpectedly fat body-derived cells were most susceptible to Cry2Ab toxin [64]. Surveys of heterologous expression of candidate targets in cell lines [65] should also take their native expression patterns into account.

8. Cry2A Toxins

An extensive sampling effort in Australia employing the F2 screen [66] yielded strains of *H. armigera* and *H. punctigera* with high levels of resistance to the Cry2Ab toxin. Linkage mapping with these strains revealed several different mutations in the ABCA2 gene that prevented expression of the full-length protein [67]. Unlike the ABCC proteins, ABCA2 has two very large ectodomains (Figure 2), and because the mutants are extremely resistant to Cry2Ab, it was speculated that the single ABCA2 protein functions similarly to the sequential binding model for the cadherin and ABCC2 [67]. Linkage mapping in a strain of *T. ni* that was selected with Dipel in British Columbia greenhouses [68] eventually resulted in the identification of a transposable element in ABCA2 conferring Cry2Ab resistance [69]. Mis-splicing mutants in ABCA2 were associated with Cry2Ab resistance in *P. gossypiella* in a laboratory-selected strain from Arizona and field populations from India [70]. Additional crosses confirmed these mutations but suggested that an additional, uncharacterized mechanism was also involved in Cry2Ab resistance in this species [71].

A different member of the ABCC family from *H. armigera* (GenBank Accession No. KY796050) was also found to bind Cry2Ab, identified by the authors as "HaABCC1" [72], although it is not the ortholog of the ABCC1 (BGIBMGA007737+38) on *B. mori* Chromosome 15 next to ABCC2 and ABCC3 described previously [24,26,28]. The ortholog of "HaABCC1" in *B. mori* is part of a small cluster on Chromosome 12 (Figure 1D) of ABCC proteins with 5 additional N-terminal transmembrane domains as well as two very large ectodomains, unlike the Chromosome 15 ABCC1-ABCC2-ABCC3 cluster in *B. mori* and other Lepidoptera (Figure 2). Although the authors speculated on the role of ABCC proteins in cross-resistance between Cry1Ac and Cry2Ab, no binding or toxicity studies were performed with Cry1Ac, and the strain of *H. armigera* used was susceptible to both Cry1Ac and Cry2Ab [72]. The use of a name already assigned to another gene has unfortunately created the false impression that the results are relevant to cross-resistance. Here we designate this gene as ABCC6 (Figures 1D and 2).

9. CRISPR/Cas9 Knockouts

CRISPR/Cas9 knockouts provide a very useful tool to investigate gene function in non-model organisms. The first use of the technique to knock out ABC transporters in Lepidoptera targeted the half-transporter genes *white*, *scarlet*, and *ok* in *Helicoverpa armigera* [73]. These are homologs of the well-known pigment transporters *white*, *scarlet*, and *brown* in *Drosophila melanogaster*, and as expected, the knockouts affected adult eye color and larval skin pigmentation. Homozygotes for the *white* knockout, however, were embryonic lethal in *H. armigera* and in the milkweed bug [74], which was unexpected because these are viable in *Drosophila*, *Aedes*, and *Tribolium*. Lethality has also complicated the interpretation for some knockouts of Cry toxin targets.

Knockouts of ABCA2 in a susceptible strain of *H. armigera* conferred >100-fold resistance to Cry2Aa and Cry2Ab, and eliminated Cry2Ab binding to BBMV, but did not affect resistance or binding to Cry1Ac [75]. After mapping Cry2Ab resistance in *T. ni* to ABCA2, where a transib mobile element was found to disrupt the gene, either ABCA1 or ABCA2 were knocked out in a susceptible strain and only ABCA2 was found to affect Cry2A tolerance [69]. Knockouts of the ABCA2 gene in *B. mori* (using the TALEN technique conferred Cry2A resistance on larvae, and heterologous expression of ABCA2 in HEK23 cells confirmed the absence of cross-resistance to Cry1A, Cry1Ca, Cry1Da, Cry1Fa, an Cry9Aa toxins [76].

In *P. xylostella*, a knockout of ABCC2 conferred 724-fold resistance to Cry1Ac and knockout of ABCC3 conferred 413-fold resistance to the same toxin. Each knockout greatl reduced BBMV binding to Cry1Ac, but the double knockout was not made in this study [77 Somewhat different results were obtained in another study of the same species [78], i which single knockouts were weakly resistant (~4-fold) and only the double knockout wa >8000-fold resistant to Cry1Ac. So far there is no explanation for the differing results in *xylostella*. A study in *H. armigera*, however, produced results closer to the second study, i that single knockouts of ABCC2 and ABCC3 were weakly resistant to Cry1Ac, while th double knockout was >15,000-fold resistant [79].

Knocking out ABCC2 in *O. furnacalis* conferred >300-fold resistance to Cry1Fa but les than 10-fold resistance to Cry1Ab or Cry1Ac and no resistance to Cry1Aa [80]. Knockin out ABCC2 in *S. frugiperda* increased tolerance to either Cry1Fa or Cry1Ab >120-fol while knocking out ABCC3 increased tolerance by a lesser amount (>16-fold); in thi study the double knockout was reported to be lethal [46]. In *S. exigua*, an ambitious stud created single knockouts of ABCC1, ABCC2, or ABCC3, as well as the cadherin and a aminopeptidase, and examined susceptibility to Cry1Ac, Cry1Fa, and Cry1Ca. Among th 15 pairwise comparisons, ABCC2 had a strong effect and the cadherin a weak effect o Cry1Ac or Cry1Fa tolerance, and ABCC2 also had a weak effect on Cry1Ca tolerance [52

CRISPR/Cas9 knockout experiments are useful in confirming the role of a given AB transporter in susceptibility to a given toxin, and when more than one knockout and mor than one toxin are compared, in assessing their relative importance. The possibility non-target effects needs more investigation, in order to reconcile studies where doubl knockouts are lethal with those where they confer even more resistance, since these studie make diametrically opposed recommendations for resistance management. Targeting th nucleotide binding domains would increase the probability of nontarget effects, sinc these are more highly conserved across ABC family members. In addition, no knock-in have been reported yet for ABC transporters, as in the case of another Bt resistance gen tetraspanin [81].

10. Negative Cross-Resistance with Chemical Insecticides

The intriguing possibility that mutations in ABC transporters could interfere wit the insect's ability to use them to rid itself of other toxins has motivated many recen studies. With the increasing use of Bt sprays and transgenic plants in the 1990s, the issue cross-resistance between Bt and chemical insecticides had achieved some attention, but E resistance mechanisms had not been characterized at the molecular level. More recently, i 2016 when several chemical insecticides were screened against an ABCC2-mutant strai of *H. armigera*, abamectin and spineotram were more toxic compared with their activit against a Cry1Ac-susceptible strain [82]. Measurements of higher abamectin concentration in mutant larvae and transfected cells were consistent with the bioassay results. RNA silencing of ABCC2 decreased susceptibility to Cry1Ac and increased susceptibility t abamectin [82]. However, the selective differential exerted by abamectin on the Cry1A resistant versus Cry1Ac-susceptible strains was small, and not all subsequent studies hav confirmed the effect.

Single and double knockouts of ABCC2 and ABCC3 in *H. armigera* produced in different study were not more susceptible to abamectin or spinetoram [79]. The knockou of ABCC2 in *O. furnacalis* was not more susceptible to abamectin or chlorantraniliprole [80 On the other hand, single knockouts of ABCC2 or ABCC3 in *S. frugiperda* were mor susceptible to abamectin and spinosad (while the double knockout was reported to b lethal and could not be compared) [46]. Another study on *S. frugiperda* found that

Cry1F-resistant strain isolated from the field with a frameshift mutation in ABCC2 had lower sensitivity to bifenthrin and higher sensitivity to spinetoram; yet when ABCC2 was knocked out in a different strain, Cry1F resistance increased 25-fold but sensitivity to chlorantraniliprole, bifenthrin, spinetoram, and acephate was unchanged [83]. Knocking out the P-glycoprotein ABCB1 in *S. exigua* increased, rather than decreased, susceptibility to abamectin and emamectin benzoate [84]; whether this protein is orthologous to the coleopteran ABCB1 (see below) has not been determined. Contradictory results were obtained in a study of ABCC2 in *P. xylostella* [85]; HEK-293 cells stably transformed with ABCC2 accumulated less avermectin, but down-regulating ABCC2 in vivo with RNAi had no effect on avermectin or chlorfenapyr tolerance.

Although results are suggestive in some studies, the changes in tolerance to the conventional insecticides examined are small and would not be useful in a resistance-breaking approach against ABCC mutations, as mortality of Bt-resistant insects would not be much greater than their Bt-susceptible counterparts. Only a few insecticides have been examined so far, and some with a greater effect are likely to be found eventually in a wider screen.

11. ABCB (P-glycoprotein) and Cry3 Toxins in Coleoptera

A different family of ABC transporters, the P-glycoproteins (ABCB), is involved in toxicity of the Cry3 toxins in Coleoptera. Linkage mapping in a strain of the poplar leaf beetle *Chrysomela tremula* (Fabricius) resistant to transgenic Cry3Aa-expressing poplar identified a frameshift in the ABCB1 protein, and heterologous expression of ABCB1 in Sf9 cells conferred susceptibility to Cry3Aa in vitro [86]. In the western corn rootworm *D. virgifera virgifera*, heterologous expression of the orthologous ABCB1 protein also conferred Cry3A sensitivity on Sf9 and HEK-293 cells in vitro, and an mCry3A-resistant strain was found to have deletions in the ABCB1 gene [87].

Whether the beetle ABCB1 genes are orthologous to PxABCB1 from *P. xylostella* mentioned in Section 3 above [39] is not known; Cry3 toxins were not experimentally tested in that study. The authors pointed out structural similarities among Cry1 and Cry3 toxins, and searched for but could not find PxABCB1 in the fragmented *Plutella* genome sequence. We have found that the *Bombyx* ortholog maps to Chromosome 15, about 1 MB away from the ABCC cluster (Figure 1B). More studies are required to confirm the involvement of the P-glycoproteins in Cry1A toxin interactions in Lepidoptera, and to establish the generality of the P-glycoproteins as targets of the beetle-active Cry3 toxins. Cross-resistance studies suggest the existence of additional, different targets in beetles [88].

12. Other ABC Transporters in Lepidoptera

Suppression of the *white* gene in *P. xylostella* by RNAi reduced Cry1Ac susceptibility but was not lethal [89]; as pointed out earlier, CRISPR/Cas9 knockouts of *white* were lethal in *H. armigera* [73]. Suppression of ABCH1 in *P. xylostella* caused larval mortality but did not affect Cry1Ac resistance [90]. A gene in *O. furnacalis* identified as ABCG was down-regulated in Cry1Ab- and Cry1Ac-resistant strains [91]; it is evidently not orthologous to either of the two genes in *P. xylostella*. No mutations in half-transporter ABCG or ABCH family genes have yet been identified in Bt-resistant Lepidoptera.

13. Hypotheses on the Mechanism of Pore Insertion

The lack of three-dimensional structures of the ABCC2 or ABCA2 proteins, the Cry toxin pore embedded in the membrane, and the toxin-binding region of the cadherin has inhibited the development of detailed hypotheses on the manner by which ABC transporters facilitate pore insertion. ABC transporters could simply be another binding site on the membrane surface, increasing the local toxin concentration and increasing the rate of pore insertion, due to a concentration effect. It has been hypothesized that ABCC2 facilitates the formation of the pre-pore oligomer, in a manner similar to the cadherin [92]. It was hypothesized that active opening and closing of the ABC transporter channel would be required to pull the pre-formed pore into the membrane [13]. This hypothesis would seem to be refuted by results with a mutant ABCC2 from *S. exigua* lacking the second nucleotide-binding domain [93], as well as engineered mutants of ABCC2 from *B. mori* lacking nucleotide-binding domains [94].

As previously pointed out [15], it is difficult to explain how evolution in *Bacillus thuringiensis* has resulted in Cry2A-type toxins that target ABCA proteins, Cry1A-type toxins that target ABCC proteins, and Cry3B-type toxins that target ABCB proteins, without invoking some fundamental property that unites these very different ABC transporters. If the shared ATP-switch mechanism powering substrate transport [4] is not such a property, then we are left without a mechanistic explanation of the pore-insertion process for the three-domain Cry proteins [13]. Many other bacterial pore-forming toxins enter the membrane with dynamic conformational changes, for example Vip3A [95] or the Membrane Attack Complex [96], or the Tc toxin [97]. Whether such a dynamic process is required for Cry toxin pore formation deserves more investigation. A reasonable hypothesis at this point is that the dynamism comes from the toxin-target interaction, not just the toxin.

14. Future Perspectives

Since the first description in 2010, mutations in ABC transporters have emerged as the most important type of mutation causing resistance to the three-domain Cry toxins of *Bacillus thuringiensis*. Yet mechanistic studies have lagged behind those on other pore-forming toxins with much more complicated structures. Why? Not enough effort has been expended on determining the three-dimensional structures that will be required for a full understanding of how the toxin interacts with membrane proteins to form a membrane pore. Since the first two structures of trypsin-activated monomers of the three-domain Cry toxins revealed their structural similarity [98,99], many more have appeared confirming that this similarity is fundamental. Recently a structure of the entire protoxin was determined [100], revealing additional domains that might potentiate pore formation in some way. The low-hanging fruit has been harvested. Lacking is a structure of any ABC transporter known to interact with a Cry toxin. Lacking is a structure of any cadherin known to interact with a Cry toxin. Lacking is a structure of the Cry toxin pore in the membrane. Without these structures, theorizing about how Bt toxins work is fantasy. Recent advances in electron cryo-microscopy (cryo-EM) have made these structures attainable. It is now time to attain them.

Funding: Preparation of this review was funded by the Max-Planck-Gesellschaft. The funder had no role in the design of the study; in the collection, analyses, or interpretation of data; in the writing of the manuscript, or in the decision to publish the results.

Data Availability Statement: NCBI GenBank: www.ncbi.nlm.nih.gov, accessed on 2 April 2021. Kaikobase: kaikobase.dna.affrc.go.jp, accessed on 2 April 2021.

Acknowledgments: I thank Linda J. Gahan for years of fruitful collaboration at Clemson University, and the U. S. National Science Foundation for finally funding our grant proposal for positional cloning in *Heliothis virescens* on the seventh submission.

Conflicts of Interest: The author declares no conflict of interest.

References

1. Holland, I.B.; Cole, S.P.C.; Kuchler, K.; Higgins, C.F. (Eds.) *ABC Proteins. From Bacteria to Man*; Academic Press: London, UK, 2003; 647p.
2. Navarro-Quiles, C.; Mateo-Bonmati, E.; Micol, J.L. ABCE Proteins: From Molecules to Development. *Front. Plant Sci.* **2018**, *9*, 1125. [CrossRef]
3. Hopfner, K.P.; Karcher, A.; Shin, D.S.; Craig, L.; Arthur, L.M.; Carney, J.P.; Tainer, J.A. Structural biology of Rad50 ATPase: ATP-driven conformational control in DNA double-strand break repair and the ABC-ATPase superfamily. *Cell* **2000**, *101*, 789–800. [CrossRef]
4. Higgins, C.F.; Linton, K.J. The ATP switch model for ABC transporters. *Nat. Struct. Mol. Biol.* **2004**, *11*, 918–926. [CrossRef] [PubMed]

5.   Adang, M.J.; Crickmore, N.; Jurat-Fuentes, J.L. Diversity of *Bacillus thuringiensis* Crystal Toxins and Mechanism of Action. In *Insect Midgut and Insecticidal Proteins*; Dhadialla, T.S., Gill, S.S., Eds.; Academic Press: London, UK, 2014; Volume 47, pp. 39–87.

6.   ISAAA. *Global Status of Commercialized of Biotech/GM Crops in 2019: Biotech Crops Drive Socio-Economic Development and Sustainable Environment in the New Frontier*; International Service for the Acquisition of Agri-Biotech Applications: Ithaca, NY, USA, 2019.

7.   Tabashnik, B.E.; Carrière, Y. Successes and Failures of Transgenic Bt Crops: Global Patterns of Field-evolved Resistance. In *Bt Resistance—Characterization and Strategies for GM Crops Producing Bacillus Thuringiensis Toxins*; Soberón, M., Gao, Y., Bravo, A., Eds.; CABI: Wallingford, UK, 2015; Volume 4, pp. 1–14.

8.   Pogue, M.G. Revised status of *Chloridea* Duncan and (Westwood), 1841, for the *Heliothis virescens* species group (Lepidoptera: Noctuidae: Heliothinae) based on morphology and three genes. *Syst. Entomol.* **2013**, *38*, 523–542. [CrossRef]

9.   Yang, C.C.; Yokoi, K.; Yamamoto, K.; Jouraku, A. An update of KAIKObase, the silkworm genome database. *Database J. Biol. Databases Curation* **2021**. [CrossRef]

10. Käll, L.; Krogh, A.; Sonnhammer, E.L.L. A combined transmembrane topology and signal peptide prediction method. *J. Mol. Biol.* **2004**, *338*, 1027–1036. [CrossRef] [PubMed]

11. Käll, L.; Krogh, A.; Sonnhammer, E.L.L. Advantages of combined transmembrane topology and signal peptide prediction—The Phobius web server. *Nucleic Acids Res.* **2007**, *35*, W429–W432. [CrossRef]

12. Dermauw, W.; Van Leeuwen, T. The ABC gene family in arthropods: Comparative genomics and role in insecticide transport and resistance. *Insect Biochem. Mol. Biol.* **2014**, *45*, 89–110. [CrossRef]

13. Heckel, D.G. Learning the ABCs of Bt: ABC transporters and insect resistance to *Bacillus thuringiensis* provide clues to a crucial step in toxin mode of action. *Pestic. Biochem. Physiol.* **2012**, *104*, 103–110. [CrossRef]

14. Heckel, D.G. Roles of ABC Proteins in the Mechanism and Management of Bt Resistance. In *Bt Resistance—Characterization and Strategies for GM Crops Producing Bacillus Thuringiensis Toxins*; Soberón, M., Gao, Y., Bravo, A., Eds.; CABI: Wallingford, UK, 2015; Volume 4, pp. 98–106.

15. Heckel, D.G. How do toxins from *Bacillus thuringiensis* kill insects? An evolutionary perspective. *Arch. Insect Biochem. Physiol.* **2020**, *104*, 21673. [CrossRef]

16. Jurat-Fuentes, J.L.; Heckel, D.G.; Ferre, J. Mechanisms of Resistance to Insecticidal Proteins from *Bacillus thuringiensis*. *Annu. Rev. Entomol.* **2021**, *66*, 121–140. [CrossRef]

17. Merzendorfer, H. ABC Transporters and Their Role in Protecting Insects from Pesticides and Their Metabolites. *Adv. Insect Physiol.* **2014**, *46*, 1–72. [CrossRef]

18. Mitsuhashi, W.; Miyamoto, K. Interaction of *Bacillus thuringiensis* Cry toxins and the insect midgut with a focus on the silkworm (*Bombyx mori*) midgut. *Biocontrol. Sci. Technol.* **2020**, *30*, 68–84. [CrossRef]

19. Pardo-López, L.; Soberón, M.; Bravo, A. *Bacillus thuringiensis* insecticidal three-domain Cry toxins: Mode of action, insect resistance and consequences for crop protection. *FEMS Microbiol. Rev.* **2013**, *37*, 3–22. [CrossRef] [PubMed]

20. Sato, R.; Adegawa, S.; Li, X.Y.; Tanaka, S.; Endo, H. Function and Role of ATP-Binding Cassette Transporters as Receptors for 3D-Cry Toxins. *Toxins* **2019**, *11*, 124. [CrossRef] [PubMed]

21. Tabashnik, B.E. ABCs of Insect Resistance to Bt. *PLoS Genet.* **2015**, *11*, e1005646. [CrossRef] [PubMed]

22. Wu, C.; Chakrabarty, S.; Jin, M.H.; Liu, K.Y.; Xiao, Y.T. Insect ATP-Binding Cassette (ABC) Transporters: Roles in Xenobiotic Detoxification and Bt Insecticidal Activity. *Int. J. Mol. Sci.* **2019**, *20*, 2829. [CrossRef] [PubMed]

23. Heckel, D.G. A return to the pore—Dissecting *Bacillus thuringiensis* toxin mode of action via voltage clamp experiments. *FEBS J.* **2016**, *283*, 4458–4461. [CrossRef] [PubMed]

24. Gahan, L.J.; Pauchet, Y.; Vogel, H.; Heckel, D.G. An ABC Transporter Mutation Is Correlated with Insect Resistance to *Bacillus thuringiensis* Cry1Ac Toxin. *PLoS Genet.* **2010**, *6*, 1001248. [CrossRef]

25. Gahan, L.J.; Gould, F.; Heckel, D.G. Identification of a gene associated with Bt resistance in *Heliothis virescens*. *Science* **2001**, *293*, 857–860. [CrossRef]

26. Atsumi, S.; Miyamoto, K.; Yamamoto, K.; Narukawa, J.; Kawai, S.; Sezutsu, H.; Kobayashi, I.; Uchino, K.; Tamura, T.; Mita, K.; et al. Single amino acid mutation in an ATP-binding cassette transporter gene causes resistance to Bt toxin Cry1Ab in the silkworm, *Bombyx mori*. *Proc. Natl. Acad. Sci. USA* **2012**, *109*, E1591–E1598. [CrossRef]

27. Tanaka, S.; Miyamoto, K.; Noda, H.; Jurat-Fuentes, J.L.; Yoshizawa, Y.; Endo, H.; Sato, R. The ATP-binding cassette transporter subfamily C member 2 in *Bombyx mori* larvae is a functional receptor for Cry toxins from *Bacillus thuringiensis*. *FEBS J.* **2013**, *280*, 1782–1794. [CrossRef]

28. Park, Y.; González-Martínez, R.M.; Navarro-Cerrillo, G.; Chakroun, M.; Kim, Y.; Ziarsolo, P.; Blanca, J.; Canizares, J.; Ferré, J.; Herrero, S. ABCC transporters mediate insect resistance to multiple Bt toxins revealed by bulk segregant analysis. *BMC Biol.* **2014**, *12*, 46. [CrossRef]

29. Baxter, S.W.; Badenes-Pérez, F.R.; Morrison, A.; Vogel, H.; Crickmore, N.; Kain, W.; Wang, P.; Heckel, D.G.; Jiggins, C.D. Parallel Evolution of *Bacillus thuringiensis* Toxin Resistance in Lepidoptera. *Genetics* **2011**, *189*, 675–679. [CrossRef]

30. Tiewsiri, K.; Wang, P. Differential alteration of two aminopeptidases N associated with resistance to *Bacillus thuringiensis* toxin Cry1Ac in cabbage looper. *Proc. Natl. Acad. Sci. USA* **2011**, *108*, 14037–14042. [CrossRef]

31. Xiao, Y.T.; Zhang, T.; Liu, C.X.; Heckel, D.G.; Li, X.C.; Tabashnik, B.E.; Wu, K.M. Mis-splicing of the ABCC2 gene linked with Bt toxin resistance in *Helicoverpa armigera*. *Sci. Rep.* **2014**, *4*, 6184. [CrossRef]

32. Coates, B.S.; Siegfried, B.D. Linkage of an ABCC transporter to a single QTL that controls *Ostrinia nubilalis* larval resistance to th Bacillus thuringiensis Cry1Fa toxin. *Insect Biochem. Mol. Biol.* **2015**, *63*, 86–96. [CrossRef] [PubMed]

33. Smith, J.L.; Farhan, Y.; Schaafsma, A.W. Practical Resistance of *Ostrinia nubilalis* (Lepidoptera: Crambidae) to Cry1F *Bacillu thuringiensis* maize discovered in Nova Scotia, Canada. *Sci. Rep.* **2019**, *9*, 18247. [CrossRef]

34. Banerjee, R.; Hasler, J.; Meagher, R.; Nagoshi, R.; Hietala, L.; Huang, F.; Narva, K.; Jurat-Fuentes, J.L. Mechanism and DNA-base detection of field-evolved resistance to transgenic Bt corn in fall armyworm (*Spodoptera frugiperda*). *Sci. Rep.* **2017**, *7*, 1087 [CrossRef]

35. Flagel, L.; Lee, Y.W.; Wanjugi, H.; Swarup, S.; Brown, A.; Wang, J.L.; Kraft, E.; Greenplate, J.; Simmons, J.; Adams, N.; et a Mutational disruption of the ABCC2 gene in fall armyworm, *Spodoptera frugiperda*, confers resistance to the Cry1Fa and Cry1A.10 insecticidal proteins. *Sci. Rep.* **2018**, *8*, 7255. [CrossRef] [PubMed]

36. Boaventura, D.; Ulrich, J.; Lueke, B.; Bolzan, A.; Okuma, D.; Gutbrod, O.; Geibel, S.; Zeng, Q.; Dourado, P.M.; Martinelli, S.; et a Molecular characterization of CryiF resistance in fall armyworm, *Spodoptera frugiperda* from Brazil. *Insect Biochem. Mol. Biol.* **202** *116*, 103280. [CrossRef]

37. Boaventura, D.; Martin, M.; Pozzebon, A.; Mota-Sanchez, D.; Nauen, R. Monitoring of Target-Site Mutations Conferring Insecticid Resistance in *Spodoptera frugiperda*. *Insects* **2020**, *11*, 545. [CrossRef]

38. Guan, F.; Zhang, J.P.; Shen, H.W.; Wang, X.L.; Padovan, A.; Walsh, T.K.; Tay, W.T.; Gordon, K.H.J.; James, W.; Czepak, C.; et a Whole-genome sequencing to detect mutations associated with resistance to insecticides and Bt proteins in *Spodoptera frugiperd Insect Sci.* **2021**, *28*. [CrossRef]

39. Zhou, J.L.; Guo, Z.J.; Kang, S.; Qin, J.Y.; Gong, L.J.; Sun, D.; Guo, L.; Zhu, L.H.; Bai, Y.; Zhang, Z.Z.; et al. Reduced expressio of the P-glycoprotein gene PxABCB1 is linked to resistance to *Bacillus thuringiensis* Cry1Ac toxin in *Plutella xylostella* (L.). *Pes Manag. Sci.* **2020**, *76*, 712–720. [CrossRef]

40. Bretschneider, A.; Heckel, D.G.; Pauchet, Y. Three toxins, two receptors, one mechanism: Mode of action of Cry1A toxins fror *Bacillus thuringiensis* in *Heliothis virescens*. *Insect Biochem. Mol. Biol.* **2016**, *76*, 109–117. [CrossRef]

41. Gómez, I.; Sánchez, J.; Miranda, R.; Bravo, A.; Soberón, M. Cadherin-like receptor binding facilitates proteolytic cleavage o helix alpha-1 in domain I and oligomer pre-pore formation of *Bacillus thuringiensis* Cry1Ab toxin. *FEBS Lett.* **2002**, *513*, 242–24 [CrossRef]

42. Ma, Y.M.; Zhang, J.F.; Xiao, Y.T.; Yang, Y.C.; Liu, C.X.; Peng, R.; Yang, Y.B.; Bravo, A.; Soberón, M.; Liu, K.Y. The Cadherin Cry1A Binding-Region is Necessary for the Cooperative Effect with ABCC2 Transporter Enhancing Insecticidal Activity of *Bacillu thuringiensis* Cry1Ac Toxin. *Toxins* **2019**, *11*, 538. [CrossRef]

43. Ren, X.L.; Jiang, W.L.; Ma, Y.J.; Hu, H.Y.; Ma, X.Y.; Ma, Y.; Li, G.Q. The *Spodoptera exigua* (Lepidoptera: Noctuidae) ABCC Mediates Cry1Ac Cytotoxicity and, in Conjunction with Cadherin, Contributes to Enhance Cry1Ca Toxicity in Sf9 Cells. *J. Econ Entomol.* **2016**, *109*, 2281–2289. [CrossRef] [PubMed]

44. Martínez-Solís, M.; Pinos, D.; Endo, H.; Portugal, L.; Sato, R.; Ferré, J.; Herrero, S.; Hernández-Martínez, P. Role of *Bacillu thuringiensis* CrylA toxins domains in the binding to the ABCC2 receptor from *Spodoptera exigua*. *Insect Biochem. Mol. Biol.* **201** *101*, 47–56. [CrossRef]

45. Zhang, J.F.; Jin, M.H.; Yang, Y.C.; Liu, L.L.; Yang, Y.B.; Gomez, I.; Bravo, A.; Soberón, M.; Xiao, Y.T.; Liu, K.Y. The Cadherin Protei Is Not Involved in Susceptibility to *Bacillus thuringiensis* Cry1Ab or Cry1Fa Toxins in *Spodoptera frugiperda*. *Toxins* **2020**, *12*, 37 [CrossRef]

46. Jin, M.H.; Yang, Y.C.; Shan, Y.X.; Chakrabarty, S.; Cheng, Y.; Soberón, M.; Bravo, A.; Liu, K.Y.; Wu, K.M.; Xiao, Y.T. Two ABC transporters are differentially involved in the toxicity of two *Bacillus thuringiensis* Cry1 toxins to the invasive crop-pest *Spodopter frugiperda* (J. E. Smith). *Pest. Manag. Sci.* **2021**, *77*, 1492–1501. [CrossRef]

47. Endo, H.; Tanaka, S.; Imamura, K.; Adegawa, S.; Kikuta, S.; Sato, R. Cry toxin specificities of insect ABCC transporters closel related to lepidopteran ABCC2 transporters. *Peptides* **2017**, *98*, 86–92. [CrossRef]

48. Stevens, T.; Song, S.S.; Bruning, J.B.; Choo, A.; Baxter, S.W. Expressing a moth ABCC2 gene in transgenic *Drosophila* cause susceptibility to Bt Cry1Ac without requiring a cadherin-like protein receptor. *Insect Biochem. Mol. Biol.* **2017**, *80*, 61–70. [CrossRef

49. Tanaka, S.; Endo, H.; Adegawa, S.; Kikuta, S.; Sato, R. Functional characterization of *Bacillus thuringiensis* Cry toxin receptor explains resistance in insects. *FEBS J.* **2016**, *283*, 4474–4490. [CrossRef]

50. Wang, J.; Zhang, H.; Wang, H.D.; Zhao, S.; Zuo, Y.; Yang, Y.H.; Wu, Y.D. Functional validation of cadherin as a receptor of Bt toxi Cry1Ac in *Helicoverpa armigera* utilizing the CRISPR/Cas9 system. *Insect Biochem. Mol. Biol.* **2016**, *76*, 11–17. [CrossRef]

51. Yang, Y.H.; Yang, Y.J.; Gao, W.Y.; Guo, J.J.; Wu, Y.H.; Wu, Y.D. Introgression of a disrupted cadherin gene enables susceptibl *Helicoverpa armigera* to obtain resistance to *Bacillus thuringiensis* toxin Cry1Ac. *Bull. Entomol. Res.* **2009**, *99*, 175–181. [CrossRef]

52. Huang, J.L.; Xu, Y.J.; Zuo, Y.Y.; Yang, Y.H.; Tabashnik, B.E.; Wu, Y.D. Evaluation of five candidate receptors for three Bt toxins i the beet armyworm using CRISPR-mediated gene knockouts. *Insect Biochem. Mol. Biol.* **2020**, *121*, 103361. [CrossRef]

53. Wang, S.H.; Kain, W.; Wang, P. *Bacillus thuringiensis* Cry1A toxins exert toxicity by multiple pathways in insects. *Insect Biochem Mol. Biol.* **2018**, *102*, 59–66. [CrossRef]

54. Tanaka, S.; Miyamoto, K.; Noda, H.; Endo, H.; Kikuta, S.; Sato, R. Single amino acid insertions in extracellular loop 2 of *Bomby mori* ABCC2 disrupt its receptor function for *Bacillus thuringiensis* Cry1Ab and Cry1Ac but not Cry1Aa toxins. *Peptides* **2016**, *7* 99–108. [CrossRef]

55. Adegawa, S.; Nakama, Y.; Endo, H.; Shinkawa, N.; Kikuta, S.; Sato, R. The domain II loops of *Bacillus thuringiensis* Cry1Aa form an overlapping interaction site for two *Bombyx mori* larvae functional receptors, ABC transporter C2 and cadherin-like receptor. *Biochim. Biophys. Acta-Proteins Proteom.* **2017**, *1865*, 220–231. [CrossRef]

56. Endo, H.; Tanaka, S.; Adegawa, S.; Ichino, F.; Tabunoki, H.; Kikuta, S.; Sato, R. Extracellular loop structures in silkworm ABCC transporters determine their specificities for *Bacillus thuringiensis* Cry toxins. *J. Biol. Chem.* **2018**, *293*, 8569–8577. [CrossRef]

57. Liu, L.L.; Chen, Z.W.; Yang, Y.C.; Xiao, Y.T.; Liu, C.X.; Ma, Y.M.; Soberón, M.; Bravo, A.; Yang, Y.B.; Liu, K.Y. A single amino acid polymorphism in ABCC2 loop 1 is responsible for differential toxicity of *Bacillus thuringiensis* Cry1Ac toxin in different *Spodoptera* (Noctuidae) species. *Insect Biochem. Mol. Biol.* **2018**, *100*, 59–65. [CrossRef]

58. Guo, Z.J.; Kang, S.; Chen, D.F.; Wu, Q.J.; Wang, S.L.; Xie, W.; Zhu, X.; Baxter, S.W.; Zhou, X.G.; Jurat-Fuentes, J.L.; et al. MAPK Signaling Pathway Alters Expression of Midgut ALP and ABCC Genes and Causes Resistance to *Bacillus thuringiensis* Cry1Ac Toxin in Diamondback Moth. *PLoS Genet.* **2015**, *11*, 1005124. [CrossRef]

59. Li, J.H.; Ma, Y.M.; Yuan, W.L.; Xiao, Y.T.; Liu, C.X.; Wang, J.; Peng, J.X.; Peng, R.; Soberón, M.; Bravo, A.; et al. FOXA transcriptional factor modulates insect susceptibility to *Bacillus thuringiensis* Cry1Ac toxin by regulating the expression of toxin-receptor ABCC2 and ABCC3 genes. *Insect Biochem. Mol. Biol.* **2017**, *88*, 1–11. [CrossRef] [PubMed]

60. Wei, W.; Pan, S.; Ma, Y.M.; Xiao, Y.T.; Yang, Y.B.; He, S.J.; Bravo, A.; Soberón, M.; Liu, K.Y. GATAe transcription factor is involved in *Bacillus thuringiensis* Cry1Ac toxin receptor gene expression inducing toxin susceptibility. *Insect Biochem. Mol. Biol.* **2020**, *118*, 103306. [CrossRef]

61. Zhu, B.; Sun, X.; Nie, X.M.; Liang, P.; Gao, X.W. MicroRNA-998-3p contributes to Cry1Ac-resistance by targeting ABCC2 in lepidopteran insects. *Insect Biochem. Mol. Biol.* **2020**, *117*, 103283. [CrossRef]

62. Knowles, B.H.; Ellar, D.J. Colloid-osmotic lysis is a general feature of the mechanism of action of *Bacillus thuringiensis* delta-endotoxins with different insect specificity. *Biochim. Biophys. Acta* **1987**, *924*, 509–518. [CrossRef]

63. Chen, Z.W.; He, F.; Xiao, Y.T.; Liu, C.X.; Li, J.H.; Yang, Y.B.; Ai, H.; Peng, J.X.; Hong, H.Z.; Liu, K.Y. Endogenous expression of a Bt toxin receptor in the Cry1Ac-susceptible insect cell line and its synergistic effect with cadherin on cytotoxicity of activated Cry1Ac. *Insect Biochem. Mol. Biol.* **2015**, *59*, 1–17. [CrossRef] [PubMed]

64. Wei, J.Z.; Liang, G.M.; Wu, K.M.; Gu, S.H.; Guo, Y.Y.; Ni, X.Z.; Li, X.C. Cytotoxicity and binding profiles of activated Cry1Ac and Cry2Ab to three insect cell lines. *Insect Sci.* **2018**, *25*, 655–666. [CrossRef]

65. Soberón, M.; Portugal, L.; Garcia-Gómez, B.I.; Sánchez, J.; Onofre, J.; Gómez, I.; Pacheco, S.; Bravo, A. Cell lines as models for the study of Cry toxins from *Bacillus thuringiensis*. *Insect Biochem. Mol. Biol.* **2018**, *93*, 66–78. [CrossRef]

66. Stodola, T.J.; Andow, D.A. F-2 screen variations and associated statistics. *J. Econ. Entomol.* **2004**, *97*, 1756–1764. [CrossRef]

67. Tay, W.T.; Mahon, R.J.; Heckel, D.G.; Walsh, T.K.; Downes, S.; James, W.J.; Lee, S.F.; Reineke, A.; Williams, A.K.; Gordon, K.H.J. Insect Resistance to *Bacillus thuringiensis* Toxin Cry2Ab Is Conferred by Mutations in an ABC Transporter Subfamily A Protein. *PLoS Genet.* **2015**, *11*, e1005534. [CrossRef]

68. Song, X.; Kain, W.; Cassidy, D.; Wang, P. Resistance to *Bacillus thuringiensis* Toxin Cry2Ab in *Trichoplusia ni* Is Conferred by a Novel Genetic Mechanism. *Appl. Environ. Microbiol.* **2015**, *81*, 5184–5195. [CrossRef]

69. Yang, X.W.; Chen, W.B.; Song, X.Z.; Ma, X.L.; Cotto-Rivera, R.O.; Kain, W.; Chu, H.N.; Chen, Y.R.; Fei, Z.J.; Wang, P. Mutation of ABC transporter ABCA2 confers resistance to Bt toxin Cry2Ab in *Trichoplusia ni*. *Insect Biochem. Mol. Biol.* **2019**, *112*, 103209. [CrossRef]

70. Mathew, L.G.; Ponnuraj, J.; Mallappa, B.; Chowdary, L.R.; Zhang, J.W.; Tay, W.T.; Walsh, T.K.; Gordon, K.H.J.; Heckel, D.G.; Downes, S.; et al. ABC transporter mis-splicing associated with resistance to Bt toxin Cry2Ab in laboratory- and field-selected pink bollworm. *Sci. Rep.* **2018**, *8*, 13531. [CrossRef] [PubMed]

71. Fabrick, J.A.; LeRoy, D.M.; Unnithan, G.C.; Yelich, A.J.; Carrière, Y.; Li, X.C.; Tabashnik, B.E. Shared and Independent Genetic Basis of Resistance to Bt Toxin Cry2Ab in Two Strains of Pink Bollworm. *Sci. Rep.* **2020**, *10*, 7988. [CrossRef]

72. Chen, L.; Wei, J.Z.; Liu, C.; Zhang, W.N.; Wang, B.J.; Niu, L.L.; Liang, G.M. Specific Binding Protein ABCC1 Is Associated With Cry2Ab Toxicity in *Helicoverpa armigera*. *Front. Physiol.* **2018**, *9*, 745. [CrossRef]

73. Khan, S.A.; Reichelt, M.; Heckel, D.G. Functional analysis of the ABCs of eye color in *Helicoverpa armigera* with CRISPR/Cas9-induced mutations. *Sci. Rep.* **2017**, *7*, 40025. [CrossRef]

74. Reding, K.; Pick, L. High-Efficiency CRISPR/Cas9 Mutagenesis of the *white* Gene in the Milkweed Bug *Oncopeltus fasciatus*. *Genetics* **2020**, *215*, 1027–1037. [CrossRef]

75. Wang, J.; Wang, H.D.; Liu, S.Y.; Liu, L.P.; Tay, W.T.; Walsh, T.K.; Yang, Y.H.; Wu, Y.D. CRISPR/Cas9 mediated genome editing of *Helicoverpa armigera* with mutations of an ABC transporter gene HaABCA2 confers resistance to *Bacillus thuringiensis* Cry2A toxins. *Insect Biochem. Mol. Biol.* **2017**, *87*, 147–153. [CrossRef]

76. Li, X.Y.; Miyamoto, K.; Takasu, Y.; Wada, S.; Iizuka, T.; Adegawa, S.; Sato, R.; Watanabe, K. ATP-Binding Cassette Subfamily a Member 2 Is a Functional Receptor for *Bacillus thuringiensis* Cry2A Toxins in *Bombyx mori*, But Not for Cry1A, Cry1C, Cry1D, Cry1F, or Cry9A Toxins. *Toxins* **2020**, *12*, 104. [CrossRef]

77. Guo, Z.J.; Sun, D.; Kang, S.; Zhou, J.L.; Gong, L.J.; Qin, J.Y.; Guo, L.; Zhu, L.H.; Bai, Y.; Luo, L.; et al. CRISPR/Cas9-mediated knockout of both the PxABCC2 and PxABCC3 genes confers high-level resistance to *Bacillus thuringiensis* Cry1Ac toxin in the diamondback moth, *Plutella xylostella* (L.). *Insect Biochem. Mol. Biol.* **2019**, *107*, 31–38. [CrossRef]

78. Liu, Z.X.; Fu, S.; Ma, X.L.; Baxter, S.W.; Vasseur, L.; Xiong, L.; Huang, Y.P.; Yang, G.; You, S.J.; You, M.S. Resistance to *Bacillus thuringiensis* Cry1Ac toxin requires mutations in two *Plutella xylostella* ATP-binding cassette transporter paralogs. *PLoS Pathog.* **2020**, *16*, e1008697. [CrossRef]
79. Wang, J.; Ma, H.H.; Zhao, S.; Huang, J.L.; Yang, Y.H.; Tabashnik, B.E.; Wu, Y.D. Functional redundancy of two ABC transporter proteins in mediating toxicity of *Bacillus thuringiensis* to cotton bollworm. *PLoS Pathog.* **2020**, *16*, e1008427. [CrossRef]
80. Wang, X.L.; Xu, Y.J.; Huang, J.L.; Jin, W.Z.; Yang, Y.H.; Wu, Y.D. CRISPR-Mediated Knockout of the ABCC2 Gene in *Ostrinia furnacalis* Confers High-Level Resistance to the *Bacillus thuringiensis* Cry1Fa Toxin. *Toxins* **2020**, *12*, 246. [CrossRef]
81. Jin, L.; Wang, J.; Guan, F.; Zhang, J.P.; Yu, S.; Liu, S.Y.; Xue, Y.Y.; Li, L.L.; Wu, S.W.; Wang, X.L.; et al. Dominant point mutation in a tetraspanin gene associated with field-evolved resistance of cotton bollworm to transgenic Bt cotton. *Proc. Natl. Acad. Sci. USA* **2018**, *115*, 11760–11765. [CrossRef]
82. Xiao, Y.T.; Liu, K.Y.; Zhang, D.D.; Gong, L.L.; He, F.; Soberón, M.; Bravo, A.; Tabashnik, B.E.; Wu, K.M. Resistance to *Bacillus thuringiensis* Mediated by an ABC Transporter Mutation Increases Susceptibility to Toxins from Other Bacteria in an Invasive Insect. *PLoS Pathog.* **2016**, *12*, e1005450. [CrossRef] [PubMed]
83. Abdelgaffar, H.; Perera, O.P.; Jurat-Fuentes, J.L. ABC transporter mutations in Cry1F-resistant fall armyworm (*Spodoptera frugiperda*) do not result in altered susceptibility to selected small molecule pesticides. *Pest. Manag. Sci.* **2021**, *77*, 949–955. [CrossRef]
84. Zuo, Y.Y.; Huang, J.L.; Wang, J.; Feng, Y.; Han, T.T.; Wu, Y.D.; Yang, Y.H. Knockout of a P-glycoprotein gene increases susceptibility to abamectin and emamectin benzoate in *Spodoptera exigua*. *Insect Mol. Biol.* **2018**, *27*, 36–45. [CrossRef]
85. Xu, J.; Wang, Z.Y.; Wang, Y.F.; Ma, H.H.; Zhu, H.; Liu, J.; Zhou, Y.; Deng, X.L.; Zhou, X.M. ABCC2 participates in the resistance of *Plutella xylostella* to chemical insecticides. *Pestic. Biochem. Physiol.* **2020**, *162*, 52–59. [CrossRef]
86. Pauchet, Y.; Bretschneider, A.; Augustin, S.; Heckel, D.G. A P-Glycoprotein Is Linked to Resistance to the *Bacillus thuringiensis* Cry3Aa Toxin in a Leaf Beetle. *Toxins* **2016**, *8*, 362. [CrossRef]
87. Niu, X.P.; Kassa, A.; Hasler, J.; Griffin, S.; Perez-Ortega, C.; Procyk, L.; Zhang, J.; Kapka-Kitzman, D.M.; Nelson, M.E.; Lu, A. Functional validation of DvABCB1 as a receptor of Cry3 toxins in western corn rootworm, *Diabrotica virgifera virgifera*. *Sci. Rep.* **2020**, *10*, 15830. [CrossRef]
88. Jakka, S.R.K.; Shrestha, R.B.; Gassmann, A.J. Broad-spectrum resistance to *Bacillus thuringiensis* toxins by western corn rootworm (*Diabrotica virgifera virgifera*). *Sci. Rep.* **2016**, *6*, 27860. [CrossRef] [PubMed]
89. Guo, Z.J.; Kang, S.; Zhu, X.; Xia, J.X.; Wu, Q.J.; Wang, S.L.; Xie, W.; Zhang, Y.J. Down-regulation of a novel ABC transporter gene (*Pxwhite*) is associated with Cry1Ac resistance in the diamondback moth, *Plutella xylostella* (L.). *Insect Biochem. Mol. Biol.* **2015**, *59*, 30–40. [CrossRef] [PubMed]
90. Guo, Z.J.; Kang, S.; Zhu, X.; Xia, J.X.; Wu, Q.J.; Wang, S.L.; Xie, W.; Zhang, Y.J. The novel ABC transporter ABCH1 is a potential target for RNAi-based insect pest control and resistance management. *Sci. Rep.* **2015**, *5*, 13728. [CrossRef]
91. Zhang, T.T.; Coates, B.S.; Wang, Y.Q.; Wang, Y.D.; Bai, S.X.; Wang, Z.Y.; He, K.L. Down-regulation of aminopeptidase N and ABC transporter subfamily G transcripts in Cry1Ab and Cry1Ac resistant Asian corn borer, *Ostrinia furnacalis* (Lepidoptera: Crambidae). *Int. J. Biol. Sci.* **2017**, *13*, 835–851. [CrossRef] [PubMed]
92. Ocelotl, J.; Sánchez, J.; Gómez, I.; Tabashnik, B.E.; Bravo, A.; Soberón, M. ABCC2 is associated with *Bacillus thuringiensis* Cry1Ac toxin oligomerization and membrane insertion in diamondback moth. *Sci. Rep.* **2017**, *7*, 2386. [CrossRef]
93. Pinos, D.; Martínez-Solís, M.; Herrero, S.; Ferré, J.; Hernández-Martínez, P. The *Spodoptera exigua* ABCC2 Acts as a Cry1A Receptor Independently of its Nucleotide Binding Domain II. *Toxins* **2019**, *11*, 172. [CrossRef]
94. Tanaka, S.; Endo, H.; Adegawa, S.; Iizuka, A.; Imamura, K.; Kikuta, S.; Sato, R. *Bombyx mori* ABC transporter C2 structures responsible for the receptor function of *Bacillus thuringiensis* Cry1Aa toxin. *Insect Biochem. Mol. Biol.* **2017**, *91*, 44–54. [CrossRef]
95. Núñez-Ramírez, R.; Huesa, J.; Bel, Y.; Ferré, J.; Casino, P.; Arias-Palomo, E. Molecular architecture and activation of the insecticidal protein Vip3Aa from *Bacillus thuringiensis*. *Nat. Commun.* **2020**, *11*, 3974. [CrossRef]
96. Bayly-Jones, C.; Bubeck, D.; Dunstone, M.A. The mystery behind membrane insertion: A review of the complement membrane attack complex. *Philos. Trans. R. Soc. B-Biol. Sci.* **2017**, *372*, 20160221. [CrossRef] [PubMed]
97. Gatsogiannis, C.; Merino, F.; Roderer, D.; Balchin, D.; Schubert, E.; Kuhlee, A.; Hayer-Hartl, M.; Raunser, S. Tc toxin activation requires unfolding and refolding of a beta-propeller. *Nature* **2018**, *563*, 209–213. [CrossRef] [PubMed]
98. Grochulski, P.; Masson, L.; Borisova, S.; Pusztai-Carey, M.; Schwartz, J.L.; Brousseau, R.; Cygler, M. *Bacillus thuringiensis* CryIA(a) insecticidal toxin—Crystal structure and channel formation. *J. Mol. Biol.* **1995**, *254*, 447–464. [CrossRef] [PubMed]
99. Li, J.D.; Carroll, J.; Ellar, D.J. Crystal structure of insecticidal delta-endotoxin from *Bacillus thuringiensis* at 2.5 A resolution. *Nature* **1991**, *353*, 815–821. [CrossRef] [PubMed]
100. Evdokimov, A.G.; Moshiri, F.; Sturman, E.J.; Rydel, T.J.; Zheng, M.Y.; Seale, J.W.; Franklin, S. Structure of the full-length insecticidal protein Cry1Ac reveals intriguing details of toxin packaging into in vivo formed crystals. *Protein Sci.* **2014**, *23*, 1491–1497. [CrossRef] [PubMed]

Article

Monitoring of Target-Site Mutations Conferring Insecticide Resistance in *Spodoptera frugiperda*

Debora Boaventura [1,2], Macarena Martin [3], Alberto Pozzebon [3], David Mota-Sanchez [4] and Ralf Nauen [2,*]

1   Institute of Crop Science and Resource Conservation, University of Bonn, 53115 Bonn, Germany; de.boaventura@hotmail.com
2   Bayer AG, Crop Science Division, R&D Pest Control, 40789 Monheim, Germany
3   Department of Agronomy, Food, Natural Resources, Animals and Environment, University of Padova, 35020 Padova, Italy; macamar_20@hotmail.com (M.M.); alberto.pozzebon@unipd.it (A.P.)
4   Department of Entomology, Michigan State University, East Lansing, MI 48824, USA; motasanc@msu.edu
*   Correspondence: ralf.nauen@bayer.com; Tel.: +49-(0)2173-38-4441

Received: 22 July 2020; Accepted: 13 August 2020; Published: 18 August 2020

Simple Summary: Fall armyworm, *Spodoptera frugiperda*, is an invasive moth species and one of the most destructive pests of maize. It is native to the Americas but recently invaded (sub)tropical regions in Africa, Asia and Oceania. Fall armyworm larvae feeding on maize plants cause substantial economic damage and are usually controlled by the application of insecticides and genetically modified (GM) maize expressing *Bacillus thuringiensis* (Bt) proteins, selectively targeting fall armyworm. It has developed resistance to many different classes of insecticides and Bt proteins as well; therefore, it is important to check field populations for the presence of mutations in target proteins conferring resistance. Here, we developed molecular diagnostic tools allowing us to test the frequency of resistance alleles in field-collected populations, either alive or preserved in alcohol. We tested 34 different populations collected on four different continents for the presence of mutations conferring resistance to common classes of insecticides and Bt proteins. We detected resistance mutations which are quite widespread, whereas others are restricted to certain geographies or even completely absent. The established molecular methods show robust results in samples collected across a broad geographical range and can be used to support decisions for sustainable fall armyworm control and applied resistance management.

Abstract: Fall armyworm (FAW), *Spodoptera frugiperda*, a major pest of corn and native to the Americas, recently invaded (sub)tropical regions worldwide. The intensive use of insecticides and the high adoption of crops expressing *Bacillus thuringiensis* (Bt) proteins has led to many cases of resistance. Target-site mutations are among the main mechanisms of resistance and monitoring their frequency is of great value for insecticide resistance management. Pyrosequencing and PCR-based allelic discrimination assays were developed and used to genotype target-site resistance alleles in 34 FAW populations from different continents. The diagnostic methods revealed a high frequency of mutations in acetylcholinesterase, conferring resistance to organophosphates and carbamates. In voltage-gated sodium channels targeted by pyrethroids, only one population from Indonesia showed a mutation. No mutations were detected in the ryanodine receptor, suggesting susceptibility to diamides. Indels in the ATP-binding cassette transporter C2 associated with Bt-resistance were observed in samples collected in Puerto Rico and Brazil. Additionally, we analyzed all samples for the presence of markers associated with two sympatric FAW host plant strains. The molecular methods established show robust results in FAW samples collected across a broad geographical range and can be used to support decisions for sustainable FAW control and applied resistance management.

Keywords: fall armyworm; insecticide resistance; target-site mutations; Bt resistance; corn strain; rice strain; resistance management; Indonesia; Kenya

1. Introduction

The fall armyworm (FAW), *Spodoptera frugiperda* (J.E. Smith) (Lepidoptera: Noctuidae), is an important agricultural pest of several crops in the western hemisphere [1,2]. Since 2016, FAW distribution expanded globally by invading different continents, first reported in Africa and later reaching Southeast Asia and, more recently, Australia, totalizing its presence in 107 countries worldwide [3–6]. The success of FAW spread is due to many factors, such as the high reproductive capacity, long-distance migration and high polyphagia [7,8].

Two sympatric host plant strains of *S. frugiperda* have been previously described: the corn strain, which feeds on large grasses such as corn and sorghum, and the rice strain, preferring small grasses such as rice [9,10]. The two strains differ not only in their host preferences but also regarding their physiology [11], insecticide susceptibility [12] and composition of genes involved in chemoreception, detoxification and digestion [13].

In South America, both strains have been already identified in field populations using molecular markers, and most populations were structured in agreement with their host preferences [14,15]. Although initial studies on the genetic structure of FAW populations from the newly invaded countries suggest a common source of origin, probably from Florida or the Caribbean [16,17], there are differences in the strain haplotypes and disagreements regarding the molecular marker and host plant that may imply inter-population movement of FAW populations from African and Asian countries [18].

An understanding of the genetic background of FAW is essential for resistance management strategies in different regions. Besides the intrinsic variation in insecticide susceptibility associated with FAW strains [11,12,19,20], the impact of migration on insecticide resistance will depend on the pre-existence of resistance alleles in the starting population and selection pressure on the newly invaded areas and spread [14,17].

At present, the Arthropod Pesticide Resistance Database (APRD) reports 144 cases of insecticide resistance in FAW globally. Among the 41 different active substances affected, 45% of the cases belong to proteins produced by *Bacillus thuringiensis* (Bt), 26% and 19% to insecticides targeting the voltage-gated sodium channel (VGSC), and acetylcholinesterase (AChE), respectively [21]. The high number of cases reported for Bt proteins, particularly those expressed in transgenic corn, reflects the intensive adoption of transgenic crops, which corresponded to 191.7 million ha worldwide in 2018 [22]. The adoption of transgenic crops expressing insect-resistant traits to control lepidopteran pests is most advanced in the United States and Brazil. Nevertheless, in Asia, the adoption of Bt-corn is high, particularly in China and India, while it is rather limited to just a few countries in Africa [22].

Many resistance cases are reported for pyrethroid insecticides targeting the VGSC and inhibitors of AChE (i.e., carbamates and organophosphates). This is due to low application costs, a high number of compounds registered for decades and frequent applications [23]. Nevertheless, together, they still account for around 30% of the global insecticide market share [23]. The most modern chemical class used to control lepidopteran pests are the diamide insecticides, acting on the ryanodine receptor (RyR) and used in different agronomic settings [23].

It is unclear whether FAW populations present in Africa were already resistant to old chemical compounds [24]. However, farmers have complained about the efficacy of pyrethroids and organophosphate insecticides under field conditions [25]. Hence, this has led to misuse by increasing rates, application frequency or even the use of unregistered compounds [25,26]. On the other hand, if no control measures against FAW are adopted, the yield losses for corn could reach up to 20.6 m tons per annum for only 12 corn-producing countries in Africa [24].

Insecticide resistance is usually conferred by the insensitivity of the target receptor and/or pharmacokinetic processes modifying the rate or the properties of the insecticides delivered to the target site [27]. Amino acid substitutions/indels at the VGSC (T929I, L932F, and L1014F), AChE (A201S,

G227A, and F290V), RyR (I4790M and G4946E) and ATP-binding cassette subfamily C2 transporter (ABCC2) (GC insertion and GY deletion) have been linked to resistance in *S. frugiperda* to pyrethroids, carbamates and organophosphates, diamides and Bt proteins (e.g., Cry1F), respectively [28–30].

In the present study, we monitored the frequency of the above-mentioned target-site mutations in the VGSC, AChE, RyR and ABCC2 in 34 populations of *S. frugiperda* collected in Brazil, Puerto Rico, Kenya and Indonesia by PCR-based allelic discrimination assays as well as pyrosequencing diagnostics. We validated and established robust diagnostic tools based on genomic DNA, which can be implemented to support decisions for appropriate resistance management strategies.

2. Materials and Methods

2.1. Insect Collection

Larvae of FAW were collected from different sites in Brazil, Puerto Rico, Kenya and Indonesia (Figure 1 and Table S1) and kept in 70% ethanol or RNAlater® (Life Technology, Carlsbad, CA, USA) until DNA extraction and genotyping.

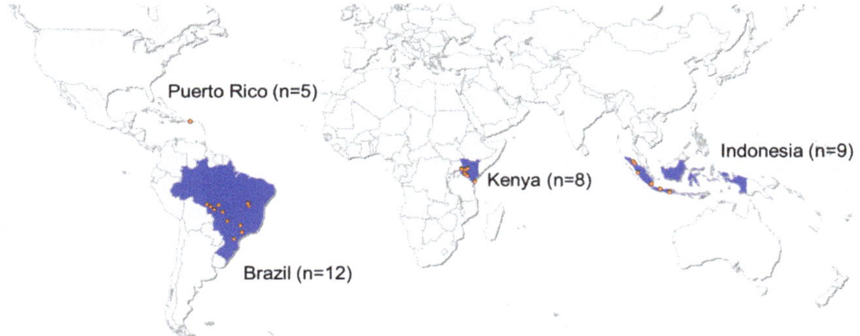

Figure 1. Map showing the origin of 34 fall armyworm populations collected in Brazil, Puerto Rico, Kenya and Indonesia (more details about collection sites in Table S1). All samples were used for the genotyping of target-site mutations. The schematic map was created using EasyMap software (Lutum + Tappert DVBeratung GmbH, Bonn, Germany).

2.2. DNA Extraction

Genomic DNA was extracted from individual larvae (whole body for second/third instar and abdominal fragments of fifth instar larvae). At least five individuals per FAW population (Table S1) were used for gDNA extraction and the total sample number used for the genotyping analysis is shown in Table 1. Agencourt DNAdvance™ (Beckmann Coulter, Beverly, CA, USA) and DNeasy Blood & Tissue Kit (QIAGEN, Hilden, Germany) were used to extract gDNA for the pyrosequencing and fluorescent probe assay, respectively. Both kits were used according to the suppliers' recommended protocols.

Table 1. Genotyping by pyrosequencing for different target-site mutations in major insecticide targets. In total, larvae of 34 populations from Brazil, Puerto Rico, Kenya and Indonesia were analyzed. Homozygous susceptible (SS), heterozygotes (RS) and homozygous resistant (RR).

Target	Country	Mutation	N	SS (%)	RS (%)	RR (%)
	Brazil		140	100.0	0.0	0.0
	Puerto Rico	L1014F	70	100.0	0.0	0.0
	Kenya		76	100.0	0.0	0.0
	Indonesia		110	98.2	1.8	0.0
	Brazil		143	100.0	0.0	0.0
Voltage-gated sodium channel (VGSC)	Puerto Rico	L932F	64	100.0	0.0	0.0
	Kenya		75	100.0	0.0	0.0
	Indonesia		88	100.0	0.0	0.0
	Brazil		143	100.0	0.0	0.0
	Puerto Rico	T929I	64	100.0	0.0	0.0
	Kenya		75	100.0	0.0	0.0
	Indonesia		88	100.0	0.0	0.0
	Brazil		147	92.5	4.1	3.4
	Puerto Rico	A201S	29	100.0	0.0	0.0
	Kenya		76	89.5	10.5	0.0
	Indonesia		85	77.6	22.4	0.0
	Brazil		127	55.1	44.9	0.0
Acetylcholinesterase (AChE)	Puerto Rico	F290V	70	4.3	10.0	85.7
	Kenya		76	26.3	47.4	26.3
	Indonesia		86	19.8	55.8	24.4
	Brazil		161	55.3	32.3	12.4
	Puerto Rico	G227A	29	100.0	0.0	0.0
	Kenya		76	100.0	0.0	0.0
	Indonesia		86	83.7	16.3	0.0
	Brazil		140	100.0	0.0	0.0
	Puerto Rico	G4946E	70	100.0	0.0	0.0
	Kenya		76	100.0	0.0	0.0
Ryanodine receptor (RyR)	Indonesia		90	100.0	0.0	0.0
	Brazil [a]		140	100.0	0.0	0.0
	Puerto Rico	I4790M	70	100.0	0.0	0.0
	Kenya		76	100.0	0.0	0.0
	Indonesia		90	100.0	0.0	0.0
	Brazil [b]		211	39.83	14.30	45.87
ATP-binding cassette transporter subfamily C (ABCC2)	Puerto Rico	GY del	19	100.0	0.0	0.0
	Kenya		70	100.0	0.0	0.0
	Indonesia		79	100.0	0.0	0.0

[a] Data published by Boaventura et al. (2020a) [30]; [b] Data published by Boaventura et al. (2020b) [28].

2.3. PCR and qPCR Conditions

PCR for pyrosequencing, PCR-RFLP and PCR for sequencing were performed in 30 µL reaction mixture containing 15 µL JumpStart™ Taq ReadyMix™ (Sigma-Aldrich, St. Louis, MO, USA), 500 nM of forward and reverse primers (Table S2), around 20 to 50 ng gDNA and nuclease-free water. The cycling conditions comprised 95 °C for 3 min, followed by 40 cycles of 95 °C for 30 s, the respective annealing temperature according to Table S2 for 30 s and 72 °C for 45 s, and a final elongation step at 72 °C for 5 min.

The fluorescent probe assays for detection of mutations F290V, I4790M and a GC insertion in the ABCC2 consisted of reactions set up at a final volume of 10 µL, with 5 µL SsoAdvanced™ Universal Probes Supermix (Bio-Rad, Hercules, CA, USA), 700 nM of forward and reverse primers (Table S2), 200 nM of probes, 20–50 ng of gDNA and nuclease-free water, and the reactions were run in duplicate. The conditions of PCR amplification were 95 °C for 5 min and 40 cycles at 95 °C for 15 s and 60 °C for 30 s. The real-time PCR was conducted in a CFX-384 real-time thermocycler (Bio-Rad, Hercules, CA, USA) and the end-point fluorescence values, taking cycle 35 as a threshold, were plotted in a scatter-plot using Bio-Rad qPCR analysis software CFX Maestro 1.0 (Bio-Rad, Hercules, CA, USA).

2.4. Characterization of S. frugiperda Strains

2.4.1. Characterization of COI Haplotypes Using PCR-RFLP

Corn and rice strain genotyping were performed using the molecular markers based on mitochondrial cytochrome oxidase subunit I (COI) with polymerase chain reaction-restriction fragment length polymorphism (PCR-RFLP), according to Nagoshi et al. (2007, 2012) [31,32]. Three to five individuals from different populations of *S. frugiperda* collected in Kenya (EP-K, KV-K, NJ-K, MJ-K and MD-K), Indonesia (WS-I, DS-I, S-I, WC-I and BC-I), Brazil (Sf_Bra, Sf_Cor, MT-PL1-2, BA-SD and PR-PG) and Puerto Rico (PR60, PR61, PR62, PR63 and PR64) (Table S1) were characterized. PCR reactions were carried out according to Section 2.3, using primer JM76 and JM77 (Table S2). After amplification, 1.0 µL of FastDigest MspI (Thermo Scientific, Vilnius, Lithuania) was added to 10 µL of each PCR reaction and incubated at 37 °C for 10 min. The PCR products were verified by an automated gel electrophoresis system, according to the AL320 method (QIAxcel DNA Screening Kit v2.0, QIAGEN, Hilden, Germany). In order to validate the results, a second PCR spanning another restriction site was performed using designed forward (891F_COI) and reverse (c1303R_COI) primers (Table S2). After the amplification, the digestion step was performed by adding EcoRV (New England Biolabs, Frankfurt, Germany), according to the manufacturer's instructions.

2.4.2. Characterization of Tpi Haplotypes Using DNA Sequencing

Plant host strain identification was additionally performed using the triosephosphate isomerase (*Tpi*) gene as a genetic marker, according to Nagoshi et al. (2019) [18]. The PCR amplification was performed according to Section 2.3, using the forward (TpiE4) and reverse (850R) primers described in Table S2. The PCR products were verified by an automated gel electrophoresis system, according to the OM500 method (QIAxcel DNA Screening Kit v2.0, QIAGEN), purified using PCR Clean-up Gel Extraction kit (Macherey-Nagel, Düren, Germany) and Sanger-sequenced by Eurofins Genomics (Cologne, Germany). The obtained *S. frugiperda Tpi* nucleotide sequences were aligned with the *Tpi* sequences for corn and rice variants according to the reference genome [13] (https: //bipaa.genouest.org/data/public/sfrudb/), using Geneious software v. 10.2.3 (Biomatters Ltd., Auckland, New Zealand).

2.5. Target-Site Resistance Diagnostics by Pyrosequencing

Amino acid substitutions in the VGSC (T929I, L932F and L1014F), AChE (A201S, G227A and F290V), RyR (I4790M and G4946E) and ABCC2 (GC insertion and GY deletion) result in resistance to pyrethroid, carbamate/organophosphate, diamide and Cry1F Bt protein, respectively. Mutation sites in the VGSC, AChE and RyR are numbered according to *Musca domestica* (GenBank X96668), *Torpedo californica* (PDB ID: 1EA5) and *Plutella xylostella* (GenBank AET09964), respectively.

A pyrosequencing based genotyping assay was designed for targeting each mutation separately and performed across 34 FAW populations (see Table S1 for details about FAW populations).

Primer pairs were designed with Assay Design Software (QIAGEN, Hilden, Germany), according to sequences deposited at the National Center for Biotechnology Information (NCBI) for FAW para-type *VGSC* (GenBank KC435025) and *ace-1* (GenBank KC435023). Primers targeting FAW *RyR* (GenBank MK226188) and *ABCC2* (GenBank KY489760) were described elsewhere [28,30], as indicated in Table S2.

The PCR conditions for pyrosequencing were performed as described in Section 2.3, using primers given in Table S2. The pyrosequencing reaction was carried out as described elsewhere [33], using a sequencing primer specific for every target-site mutation analyzed, according to Table S2.

2.6. Fluorescence Based Allelic Discrimination Assays

2.6.1. F290V Mutation in AChE

Primers were designed using the OligoArchitect™ Assay Design (Sigma-Aldrich, St. Louis, MO, USA) for the detection of the F290V mutation in ace1. Allele-specific probes were labeled with FAM (Sf_F290_FAM) or HEX (Sf_F290_mut_HEX) at the 5′ end for the detection of the wildtype and mutant allele, respectively (Table S2). Five individuals from populations from Brazil (Sf_Bra, MT-PL1 and PR-PG), Puerto Rico (PR60, PR61, PR62 and PR63), Kenya (EP-K, KV-K, MJ-K, KF-K and NW-K) and Indonesia (WS-I, DS-I, S-I, WC-I and JL-I) were tested (Table S1). PCR reactions and allele discrimination analysis were performed as described in Section 2.3.

2.6.2. GC Insertion in ABCC2

The GC insertion in ABCC2 was detected according to Banerjee et al. (2017) [34], with slight modifications. Briefly, reactions were composed of a HEX-labeled probe (SfABCC2mut allele) that is *SfABCC2* mutant allele-specific and a FAM-labeled probe (SfABCC2), specific to the *SfABCC2* wildtype allele, gDNA (around 50 ng), the forward (Sf_ABCC2_F) and the reverse (Sf_ABCC2_R) primers (Table S2). The populations tested were the same as described in Section 2.6.1 and PCR reactions were prepared as mentioned in Section 2.3.

2.6.3. I4790M Mutation in the RyR

The detection of the RyR I4790M mutation was performed as described by Boaventura et al. (2020) [30] using forward (Sf_taq_I4790_F) and reverse (Sf_taq_I4790_R) primers, the mutant allele-specific FAM-labeled probe (Sf_I4790_mut_FAM) and a HEX-labeled probe (Sf_I4790_HEX) that is wildtype allele-specific (Table S2). Individuals with known genotype from strain Chlorant-R (homozygote for M4790) as well as artificial heterozygotes (a mixture of gDNA from Chlorant-R and Sf_Bra individuals) were used as internal controls. The assay was validated with populations collected in Brazil, Puerto Rico, Kenya and Indonesia, as described above (Section 2.6.1).

3. Results

3.1. Characterization of S. frugiperda Strains

The mitochondrial COI and nuclear Tpi molecular markers were employed for the identification of sympatric FAW rice and corn strain according to Nagoshi et al. [17,31,32]. The amplification of the respective COI fragment resulted in a PCR product of around 569 bp for both strains, but the fragment amplified from corn strain contained a MspI restriction site; therefore, after digestion, the PCR product was cut into two fragments (approximately 487 and 72 bp) (Figure 2A and Figure S1A). According to this method, all samples tested from Brazil and Puerto Rico were characterized as corn strain, whereas in Kenya and Indonesia, most of the individuals were characterized as rice strain, i.e., 70% and 91.7%, respectively (Figure 2A).

On the other hand, when using the EcoRV restriction site, the rice strain fragment was cut into two bands of around 350 and 150 bp and the corn strain fragment of 500 bp remained uncut (Figure 2B and Figure S1B). Again, all samples tested from Brazil and Puerto Rico were corn strain and most of the samples from Kenya and Indonesia were rice strain, i.e., 61 and 92%, respectively.

Host plant strain characterization using the *Tpi* gene was performed according to Nagoshi et al. (2019) [17]. The genetic markers used in this method are all single nucleotide substitutions present in the *TpiE4* exon. When using the primers 412F and 850R (Table S2), most of the *TpiE4* exon is amplified, producing a fragment of about 199 bp. Trimmed sequences were deposited in NCBI (GenBank MT706015–MT706018). The strain is defined by the gTpi183Y site, where the corn strain has a cytosine and the rice strain a thymine. All the samples tested from the four countries were corn strain, having a cytosine at the position gTpi183Y. It is worth mentioning that, at the position gTpi192Y, an adenine or thymine was observed in some samples from Brazil, Puerto Rico and Kenya (1, 1 and 2, respectively), while Nagoshi et al. (2019) [17] reported only a cytosine or a thymine at position Tpi192Y.

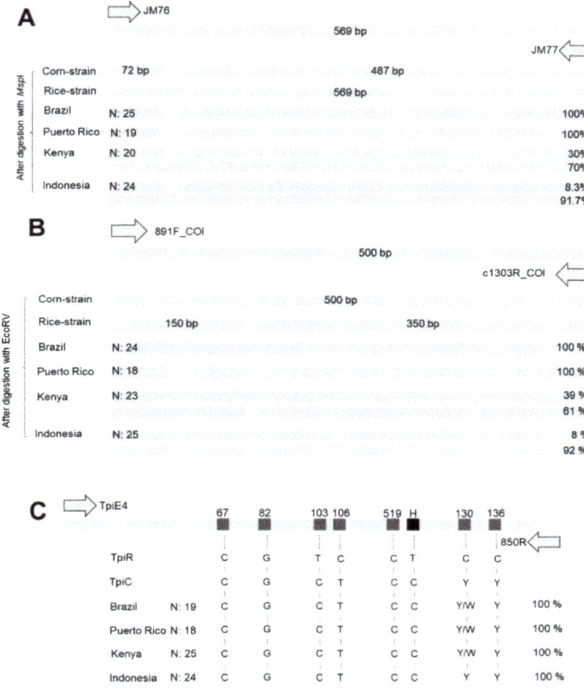

Figure 2. Schematic representation of amplified *COI* and *TpiE4* fragments used for *Spodoptera frugiperda* host plant strain identification in field samples collected in Brazil, Puerto Rico, Kenya and Indonesia. *COI* polymorphism in *S. frugiperda* was determined by RFLP-PCR. (**A**) PCR product containing a MspI restriction site in the corn strain and PCR fragments obtained after digestion with FastDigest MspI. (**B**) PCR product that contains an EcoRV strain-specific site. After digestion with EcoRV, the corn strain remains uncut, whereas the rice strain is cut. (**C**) *TpiE4* fragment with different polymorphic sites was Sanger-sequenced. Position marked H defines whether it is a rice strain (thymine; TpiR) or a corn strain (cytosine; TpiC).

3.2. Detection of Target-Site Mutations by Pyrosequencing

The pyrosequencing assay used to genotype the mutations in the VGSC revealed that almost all analyzed larvae (n = 396) were wildtype, with no mutations at those sites analyzed. Only strain NB-KA from Indonesia included a few individuals heterozygous for the L1014F mutation, corresponding to 1.8% of all samples analyzed from Indonesia (Table 1). On the other hand, the mutations T929I and L932F were not detected at all in any population tested (Table 1), suggesting the lack of target-site resistance to pyrethroids in almost all samples analyzed. Resistant AChE alleles were found at much higher frequencies across countries in many populations analyzed. The mutation F290V was detected at the highest frequency (Table 1). In Brazil, 45% of the samples genotyped (57 out of 127 larvae) were heterozygote, whereas most samples from Puerto Rico (except strain PR65) were homozygote for V290, representing 85.7% of all samples tested (60 out of 70 larvae). Populations collected in Kenya and Indonesia also carried the F290V mutation in AChE and, on average, 47% and 56% of the samples were heterozygotes, respectively. The other AChE mutation sites analyzed, A201S and G227A, were not detected in Puerto Rico, while G227A was absent also in Kenya. RyR mutations G4946E and

I4790M, conferring resistance to diamide insecticides, were not detected in any of the populations tested (Table 1); all individuals tested were homozygous wildtype at both positions. We also tested 379 individuals for the presence of a GY deletion in ABCC2, known to confer resistance to Cry1F in FAW [28]. This functionally validated target-site mutation was absent in samples collected in Puerto Rico, Kenya and Indonesia but detected in many of the tested Brazilian larvae, as recently described [28] (Table 1).

3.3. Detection of Target-Site Mutations by Fluorescence Based Allelic Discrimination Assays

As the target-site, mutation F290V in AChE was the most frequent mutation found in all populations tested. We decided to develop a PCR-based allelic discrimination assay using fluorescent probes, which could be performed at larger-scale worldwide using a qPCR machine, because pyrosequencing-based diagnosis is more expensive and less common. All larvae analyzed from Puerto Rico were homozygote for the V290 resistance allele (Figure 3A). In Kenya and Indonesia, five populations were tested and, on average, 40% and 56% of the larvae were heterozygotes, respectively (Figure 3C,D). Individuals from the two field populations from Brazil (MT-PL1 and PR-PG) were homozygous for the V290 resistance allele (Figure 3B). The population Sf_Bra was kept for 15 years under laboratory conditions without insecticide exposure and all larvae were homozygotes for the susceptible wildtype allele F290 (Figure 3B).

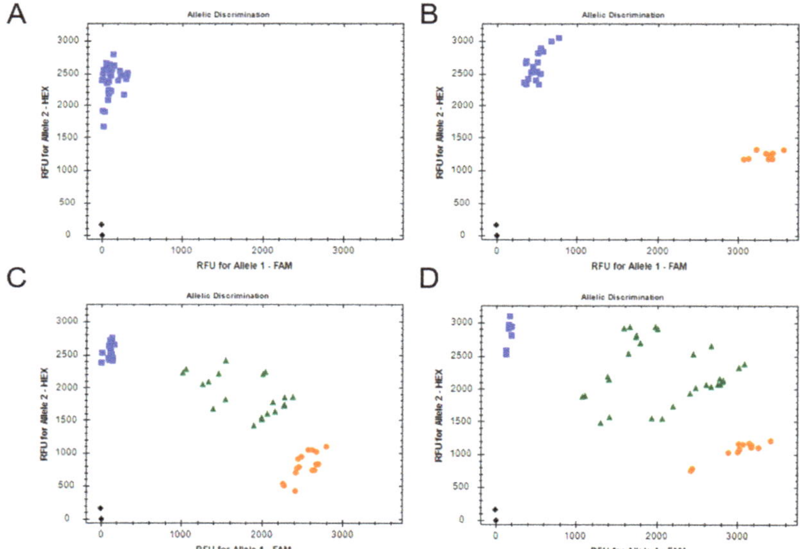

Figure 3. Bivariate plot showing the discrimination of different acetylcholinesterase alleles in *Spodoptera frugiperda* samples by an allele-specific real-time PCR fluorescent probe assay. Each dot represents a single larva. Blue squares represent mutant RR homozygotes (V290; allele 1), orange circles susceptible SS homozygotes (F290; allele 2) and green triangles SR heterozygotes (F290/V290). Analysis of fall armyworm field samples collected in (**A**) Puerto Rico, (**B**) Brazil, (**C**) Kenya and (**D**) Indonesia.

The I4790M mutation in the RyR was assessed using Chlorant-R resistant FAW larvae as a positive control for M4790. The resistant allele was not present in any other sample analyzed (Figure 4). For the detection of GC insertion at the ABCC2 causing resistance to Cry1F protein in *S. frugiperda* in Puerto Rico, the assay described by Banerjee et al. (2017) [34] was used, however substituting the VIC fluorescent probe with FAM, as described in Section 2.6.2. The GC insertion was only observed in Puerto Rican samples (Figure S2).

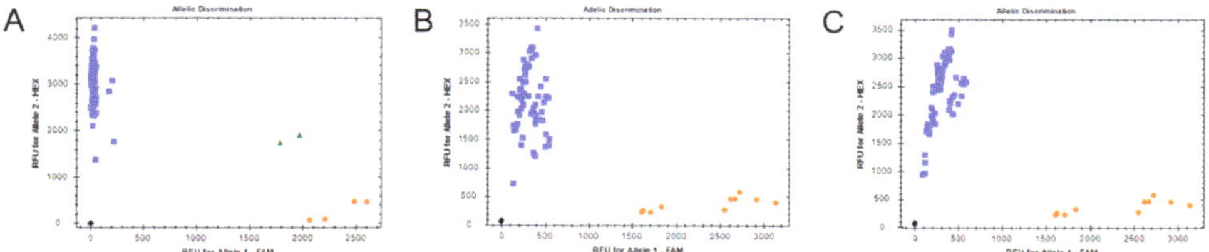

Figure 4. Detection of the RyR I4790M mutation using an allele-specific real-time PCR fluorescent probe assay, as recently described by Boaventura et al. (2020). (**A**) Genotyping of *Spodoptera frugiperda* collected in Puerto Rico represented by blue squares (wildtype SS homozygotes, I4790 allele), orange circles represent strain Chlorant-R mutant RR homozygotes from Brazil (M4790; allele 1) and green triangles artificial SR heterozygotes (I4790/M4790). All individuals tested from (**B**) Kenya and (**C**) Indonesia were susceptible homozygotes for I4790 (blue squares).

4. Discussion

The highly invasive nature and the potential economic impact of FAW have raised a lot of concerns across continents. Changes in agricultural practices and biological control are among a diverse range of measures implemented in recently invaded African countries and India by smallholder farmers and at rather low FAW infestation levels [25,26,35–39]. In countries with significant agricultural input subsidy programs, synthetic insecticides have been used to control FAW outbreaks [25,26,38]. However, in some countries, farmers claimed rather low efficacy of some of the insecticide classes used, such as organophosphates and pyrethroids [25,35]. It remains unclear whether the low field efficacy of insecticides against FAW in Africa is due to resistance or poor application technology affecting plant coverage.

Our genotyping study was conducted to shed some light on the presence of target-site insecticide resistance mechanisms in 34 populations collected in Kenya, Indonesia, Puerto Rico and Brazil. Our results from FAW populations collected in Kenya showed a relatively high frequency of the F290V mutation in AChE, the target of organophosphate and carbamate insecticides. The chance that alleles conferring resistance to these rather old chemical classes were already present at high frequency in the invasive population is quite high. Their frequency was likely augmented by further selection, using applications of cheap products based on organophosphate and carbamate chemistries. Similar findings have been reported for the tomato leafminer (*Tuta absoluta*) in Iran, where this pest has been recently introduced. Resistance to pyrethroids and organophosphates was expected in the invading populations and this expectation was supported by the identification of target-site mutations in VGSC and ace1, respectively [40].

As a result of frequent insecticide applications, multiple resistance cases have been described for field populations in regions where FAW is native [41,42]. In Brazil and Puerto Rico, for instance, resistance has been reported to pyrethroids, organophosphates, carbamates, spinosyns, benzoylureas and, most recently, diamides [29,42–47]. The genetic inheritance of insecticide resistance in FAW has been investigated, and cases of FAW resistance were described as polygenic and metabolic [46–48].

Pyrethroid insecticides are supposed to bind in the domain IIS4-S5 linker and domain IIIS6 of *para*-type sodium channels [49] and the common L1014F mutation has been reported to confer pyrethroid resistance ratios of 10-20-fold [29,50–52]. More than 30 unique resistance-associated mutations including L1014F or combinations thereof have been described in VGSC in many other different species [53]. Three mutations (T929I, L932F and L1014F) at the VGSC have been recently described in pyrethroid-resistant *S. frugiperda* from Brazil [29] and one of the mutations, L932F, was detected in FAW populations from China [54]. Our genotyping results revealed the absence of the L1014F mutation in almost all analyzed samples, except for one population from Indonesia (K-I, Table S1), where only two heterozygotes out of 30 individuals were detected. No other mutation conferring pyrethroid resistance and described for *S. frugiperda* was detected in the populations tested.

However, other mechanisms such as enhanced metabolism by elevated levels of cytochrome P450 monooxygenases are known to confer pyrethroid resistance in FAW [29] but were not tested in our study as we used gDNA of alcohol preserved FAW samples as we did not have access to living insects.

Organophosphates and carbamates target AChE and resistance is often associated with mutations in the *ace-1* gene, leading to amino acid substitutions at the enzyme's active site [55]. Our genotyping results confirmed the presence of the following amino acid substitutions A201S, G227A and F290V in populations collected in Brazil, as described by Carvalho et al. (2013) [29]. The detected point mutations co-exist, at least in heterozygous individuals, in populations BA-SD, PR-PG and MT-PL1-2. Moreover, we also detected the F290V mutation in samples from Puerto Rico, Indonesia and Kenya. Point mutations linked to organophosphate resistance have been described for *Cydia pomonella* (F399V), *Chilo suppressalis* (A314S) and *P. xylostella* (D131G, A201S, G227A and A441G) [56–60]. Moreover, heterologous expression of AChE mutants (A303S, G329A and L554S) from the silkworm (*Bombyx mori*) have supported the reduction in AChE sensitivity towards carbamate and organophosphate insecticides [61].

Diamide insecticides comprised two chemotypes, the phthalic (flubendiamide) and anthranilic acid diamides (e.g., chlorantraniliprole), which were shown to be affected differently by the presence of point mutations leading to amino acid substitutions, particularly G4946E and I4790M in the lepidopteran RyR (numbering according to the *P. xylostella* RyR)—recently reviewed by Richardson et al. [62]. Phthalic diamides are less potent against pests carrying a methionine at position 4790 [63]. In Puerto Rico, resistance ratios of 160 to 500- fold have been reported to chlorantraniliprole and flubendiamide, respectively, but the mechanisms of resistance were not studied in detail [42]. However, in our genotyping assays, G4946E and I4790M were not detected in samples from Puerto Rico or any other country. So far, the I4790M mutation in *S. frugiperda* has been detected only in one FAW strain (Chlorant-R) from Brazil, selected with chlorantraniliprole under laboratory conditions and showing resistance ratios of >230 and >42,000-fold against chlorantraniliprole and flubendiamide, respectively [30,43]. Our genotyping data suggest that the frequency of those resistance alleles (G4946E and I4790M) is low under field conditions in those locations here investigated.

Mutations in the ABCC2 transporter have been associated with Cry1F and Cry1A.105 resistance in FAW: a GC insertion causing a premature stop codon has been found in Cry1F-resistant FAW strains from Puerto Rico [34,64], whereas a functionally validated GY deletion was very recently described in Cry1F-resistant populations from Brazil [28]. While we identified the described GC insertion and GY deletion in many samples from Puerto Rico and Brazil, respectively, none of the above-mentioned mutations were found in populations from Kenya or Indonesia, supporting the absence/very low frequency of these mutations in the field and a lack of selection pressure by transgenic corn expressing Cry1F in those countries. Recent whole-genome sequencing of FAW samples collected in China, Malawi, Uganda and Brazil revealed a novel ABCC2 resistance allele in FAW collected in Brazil, leading to a truncated and likely non-functional protein [65].

Two sympatric host plant strains of *S. frugiperda* have been previously described: the corn strain and the rice strain, which prefers forage grasses and rice [9–11]. Recent studies have reported that *S. frugiperda* populations present in Asia and Africa are an inter-strain hybrid, with the genetic background mostly from the corn strain [54,66]. Therefore, we were interested in the host-plant strain composition of our samples and analyzed individual larvae of different populations using recently described markers by RFLP and PCR. Our results using COI and Tpi genetic markers confirmed that the corn strain is the most abundant in Brazil and Puerto Rico, as shown in previous studies [14,15,67,68]. The COI genetic marker used in this study revealed dominance of the rice strain in those populations that we collected in Kenya (70%, though collected from corn plants) and Indonesia (91.7%). However, to avoid any identification bias, we used a second marker, *Tpi*, and the obtained data revealed that all samples, including those from Kenya and Indonesia, resemble corn strain and none rice strain. This discrepancy between COI and *Tpi* markers has already been noticed by other authors, especially with samples from Africa and Asia, where strain characterization is dependent on the molecular marker used [16,18,65]. However, the exclusive identification of the corn strain in our samples by using the *Tpi* marker is in accordance with the preferred host, suggesting that it is a more accurate strain marker than COI. In terms of insecticide susceptibility, there is not much difference between host-plant strains, at least when considering the efficacy of recommended label rates of many insecticides. A slightly higher level of cytochrome P450 activity in corn-adopted FAW [9] may render the corn strain slightly more tolerant, but this is unlikely to result in reduced efficacy of insecticides at their recommended label rates in the absence of resistance. The rice strain has been reported to be more susceptible to diazinon, carbaryl and Bt toxins, whereas corn strain larvae were shown to be more susceptible to the carbamate carbofuran [12,19,20]. Recently, Arias et al. (2019) [14] have tested the possible influence caused by the migration of individuals from hot spots—characterized by higher LC_{50} values against flubendiamide and lufenuron. The authors concluded that migration did not play the key role but, rather, the pest management measures adopted and cropping strategies in the respective region. Therefore, we want to reinforce that, although high frequencies of alleles conferring resistance to organophosphate and carbamates were detected, the choice of the appropriate management strategy to be adopted based on regionally registered insecticides and alternative measures is likely to be the key factor for sustainable FAW control. The practical relevance of the presence of alleles conferring resistance is determined by the selection pressure adopted in the field and whether the mutations present carry any fitness cost. The resistance alleles might decrease in frequency in the absence of selection pressure or increase when the application of specific insecticides increases [14]. Therefore, strategies to slow down the development of insecticide resistance should be driven by the application of insecticides with different modes of action [14,69]. Compounds such as diamides, emamectin benzoate and spinosyns [25,70–72] have mostly shown good control of several lepidopteran pests and would be valuable tools in FAW resistance management strategies in the newly invaded countries.

5. Conclusions

Based on our genotyping results described in this study, the field efficacy of organophosphate and carbamate insecticides is likely to be compromised by the presence of the AChE V290 allele in hetero- and homozygous form in Brazil, Kenya, Indonesia and Puerto Rico. To achieve successful integrated pest management of FAW and reduce the risk of economic losses, resistance management strategies will need to be implemented at regional levels in the newly invaded countries and can be supported by using the presented diagnostic tools to detect and monitor the early presence of resistance alleles in the field.

Supplementary Materials: The following are available online at http://www.mdpi.com/2075-4450/11/8/545/s1, Table S1: Populations of *Spodoptera frugiperda* collected in different countries and years used for genotyping of target-site mutations, Table S2: List of primers for pyrosequencing and dual fluorescence probe assay used for the identification of different target-site mutations and Spodoptera frugiperda strain identification by RFLP-PCR and Sanger sequencing, Figure S1: Automated analysis of DNA fragments showing COI polymorphism in Spodoptera frugiperda. (A) PCR product containing a strain specific MspI site that was amplified using the JM76 and JM77 primers (Table S2) followed by products obtained after the digestion with FastDigest MspI. Corn-strain is cut and rice-strain remains uncut as it does not have the MspI site. (B) PCR product amplified with the primers 891F_COI and c1303R_COI (Table S2) that contains a EcoRV strain specific site. After digestion with EcoRV the corn-strain amplicon remains uncut whereas it is cut in the rice-strain. Details about samples, see Table S1, Figure S2: Detection of GC insertion allele at the ATP-binding cassette subfamily C2 (ABCC2) conferring resistance to Bacillus thuringiensis Cry1F toxin using PCR fluorescent probe assay described by Banerjee et al [34]; Blue squares represent mutant ABCC2 homozygotes for the GC insertion, orange circles ABCC2 wildtype SS homozygotes, and green triangles SR representing heterozygotes. Analysis of fall armyworm field samples collected in (A) Brazil, (B) Puerto Rico, (C) Kenya, and (D) Indonesia.

Author Contributions: R.N. and D.B. conceived the study. R.N. supervised the project. D.B. and M.M. performed the research and analyzed data. D.M.-S. supported the study with biological material. A.P. supervised M.M. and acquired funding. D.B. and R.N. wrote the paper. All authors have read and agreed to the published version of the manuscript.

Funding: M.M. thanks the Erasmus+–KA1 Erasmus Mundus Joint Master's Degree Programme of the European Commission as this paper has been developed as a result of a mobility stay funded by the PLANT HEALTH Project.

Acknowledgments: The authors thank numerous field researchers from Bayer for collecting FAW samples in Brazil, Kenya and Indonesia.

Conflicts of Interest: R.N. is employed by Bayer AG, a manufacturer of insecticides and transgenic maize. This work was partially financed by Bayer AG. The authors declare no additional conflicts of interest.

References

1.   Luginbill, P. The fall armyworm. *U.S. Dept. Agric. Tech. Bull.* **1928**, *34*, 1–91.
2.   Pogue, M.G. A world revision of the genus *Spodoptera Gueneé* (Lepidoptera: Noctuidae). *Mem. Am. Ent. Soc.* **2002**, *43*, 1–202.
3.   EPPO. Available online: https://gd.eppo.int/taxon/LAPHFR/distribution (accessed on 25 March 2020).
4.   Goergen, G.; Kumar, P.L.; Sankung, S.B.; Togola, A.; Tamò, M. First Report of Outbreaks of the Fall Armyworm *Spodoptera frugiperda* (J E Smith) (Lepidoptera, Noctuidae), a new alien invasive pest in west and central Africa. *PLoS ONE* **2016**, *11*, e0165632. [CrossRef] [PubMed]
5.   Kalleshwaraswamy, C.M.; Asokan, R.; Swamv, H.M.M.; Marutid, M.S.; Pavithra, H.B.; Hegde', K.; Navi', S.; Prabhu', S.T.; Goergen, G. First report of the Fall armyworm, *Spodoptera frugiperda* (J E Smith) (Lepidoptera: Noctuidae), an alien invasive pest on maize in India. *Pest Manag. Hortic. Ecosyst.* **2018**, *24*, 23–29.
6.   Shylesha, A.N.; Jalali, S.K.; Gupta, A.; Varshney, R.; Venkatesan, T.; Shetty, P.; Ojha, R.; Ganiger, P.C.; Navik, O.; Subaharan, K.; et al. Studies on new invasive pest *Spodoptera frugiperda* (J E Smith) (Lepidoptera: Noctuidae) and its natural enemies. *Biol. Control* **2018**, *32*, 145–151. [CrossRef]
7.   Barros, E.M.; Torres, J.B.; Ruberson, J.R.; Oliveira, M.D. Development of *Spodoptera frugiperda* on different hosts and damage to reproductive structures in cotton: Fall armyworm performance on different hosts. *Entomol. Exp. Appl.* **2010**, *137*, 237–245. [CrossRef]
8.   Westbrook, J.K.; Nagoshi, R.N.; Meagher, R.L.; Fleischer, S.J.; Jairam, S. Modeling seasonal migration of fall armyworm moths. *Int. J. Biometeorol.* **2016**, *60*, 255–267. [CrossRef]
9.   Prowell, D.P.; McMichael, M.; Silvain, J.-F. Multilocus Genetic Analysis of Host Use, Introgression, and Speciation in Host Strains of Fall Armyworm (Lepidoptera: Noctuidae). *Ann. Entomol. Soc. Am.* **2004**, *97*, 1034–1044. [CrossRef]
10.   Veenstra, K.H.; Pashley, D.P.; Ottea, J.A. Host-Plant Adaptation in Fall Armyworm Host Strains: Comparison of Food Consumption, Utilization, and Detoxication Enzyme Activities. *Ann. Entomol. Soc. Am.* **1995**, *88*, 80–91. [CrossRef]
11.   Pashley, D.P. Quantitative genetics, development, and physiological adaptation in host strains of fall armyworm. *Evolution* **1988**, *42*, 93–102. [CrossRef]
12.   Adamczyk, J.J.; Leonard, B.R.; Graves, J.B. Toxicity of Selected Insecticides to Fall Armyworms (Lepidoptera: Noctuidae) in Laboratory Bioassay Studies. *Fla. Entomol.* **1999**, *82*, 230–236. [CrossRef]
13.   Gouin, A.; Bretaudeau, A.; Nam, K.; Gimenez, S.; Aury, J.-M.; Duvic, B.; Hilliou, F.; Durand, N.; Montagné, N.; Darboux, I.; et al. Two genomes of highly polyphagous lepidopteran pests (*Spodoptera frugiperda*, Noctuidae) with different host-plant ranges. *Sci. Rep.* **2017**, *7*, 11816. [CrossRef] [PubMed]
14.   Arias, O.; Cordeiro, E.; Corrêa, A.S.; Domingues, F.A.; Guidolin, A.S.; Omoto, C. Population genetic structure and demographic history of *Spodoptera frugiperda* (Lepidoptera: Noctuidae): Implications for insect resistance management programs. *Pest Manag. Sci.* **2019**, *75*, 2948–2957. [CrossRef] [PubMed]
15.   Silva-Brandão, K.L.; Peruchi, A.; Seraphim, N.; Murad, N.F.; Carvalho, R.A.; Farias, J.R.; Omoto, C.; Cônsoli, F.L.; Figueira, A.; Brandão, M.M. Loci under selection and markers associated with host plant and host-related strains shape the genetic structure of Brazilian populations of *Spodoptera frugiperda* (Lepidoptera, Noctuidae). *PLoS ONE* **2018**, *13*, e0197378. [CrossRef] [PubMed]
16.   Nagoshi, R.N.; Htain, N.N.; Boughton, D.; Zhang, L.; Xiao, Y.; Nagoshi, B.Y.; Mota-Sanchez, D. Southeastern Asia fall armyworms are closely related to populations in Africa and India, consistent with common origin and recent migration. *Sci. Rep.* **2020**, *10*, 1421. [CrossRef]
17.   Nagoshi, R.N.; Dhanani, I.; Asokan, R.; Mahadevaswamy, H.M.; Kalleshwaraswamy, C.M.; Sharanabasappa; Meagher, R.L. Genetic characterization of fall armyworm infesting South Africa and India indicate recent introduction from a common source population. *PLoS ONE* **2019**, *14*, e0217755. [CrossRef]

18. Nagoshi, R.N.; Goergen, G.; Plessis, H.D.; Van den Berg, J.; Meagher, R. Genetic comparisons of fall armyworm populations from 11 countries spanning sub-Saharan Africa provide insights into strain composition and migratory behaviors. *Sci. Rep.* **2019**, *9*, 8311. [CrossRef]
19. Ingber, D.A.; Mason, C.E.; Flexner, L. Cry1 Bt Susceptibilities of Fall Armyworm (Lepidoptera: Noctuidae) Host Strains. *J. Econ. Entomol.* **2018**, *111*, 361–368. [CrossRef]
20. Pashley, D.P.; Sparks, T.C.; Quisenberry, S.S.; Jamjanya, T. Dowd. Two fall armyworm strains feed on corn, rice and Bermuda-grass. *La. Agric.* **1987**, *30*, 8–9.
21. APRD. Available online: https://www.pesticideresistance.org/ (accessed on 23 March 2020).
22. ISAAA. Available online: http://www.argenbio.org/adc/uploads/ISAAA_2017/isaaa-brief-53-2017 (accessed on 23 March 2020).
23. Sparks, T.C.; Nauen, R. IRAC: Mode of action classification and insecticide resistance management. *Pestic. Biochem. Physiol.* **2015**, *121*, 122–128. [CrossRef]
24. Day, R.; Abrahams, P.; Bateman, M.; Beale, T.; Clottey, V.; Cock, M.; Colmenarez, Y.; Corniani, N.; Early, R.; Godwin, J.; et al. Fall Armyworm: Impacts and implications for Africa. *Outlooks Pest manag.* **2017**, *28*, 196–201. [CrossRef]
25. Kumela, T.; Simiyu, J.; Sisay, B.; Likhayo, P.; Mendesil, E.; Gohole, L.; Tefera, T. Farmers' knowledge, perceptions, and management practices of the new invasive pest, fall armyworm (*Spodoptera frugiperda*) in Ethiopia and Kenya. *Int. J. Pest Manag.* **2019**, *65*, 1–9. [CrossRef]
26. Sisay, B.; Tefera, T.; Wakgari, M.; Ayalew, G.; Mendesil, E. The Efficacy of Selected Synthetic Insecticides and Botanicals against Fall Armyworm, *Spodoptera frugiperda*, in Maize. *Insects* **2019**, *10*, 45. [CrossRef] [PubMed]
27. Feyereisen, R. Molecular biology of insecticide resistance. *Toxicol. Lett.* **1995**, 83–90. [CrossRef]
28. Boaventura, D.; Ulrich, J.; Lueke, B.; Bolzan, A.; Okuma, D.; Gutbrod, O.; Geibel, S.; Zeng, Q.; Dourado, P.M.; Martinelli, S.; et al. Molecular characterization of Cry1F resistance in fall armyworm, *Spodoptera frugiperda* from Brazil. *Insect Biochem. Mol. Biol.* **2020**, *116*, 103280. [CrossRef]
29. Carvalho, R.A.; Omoto, C.; Field, L.M.; Williamson, M.S.; Bass, C. Investigating the Molecular Mechanisms of Organophosphate and Pyrethroid Resistance in the Fall Armyworm *Spodoptera frugiperda*. *PLoS ONE* **2013**, *8*, e62268. [CrossRef]
30. Boaventura, D.; Bolzan, A.; Padovez, F.E.; Okuma, D.M.; Omoto, C.; Nauen, R. Detection of a ryanodine receptor target-site mutation in diamide insecticide resistant fall armyworm, *Spodoptera frugiperda*. *Pest. Manag. Sci.* **2020**, *76*, 47–54. [CrossRef]
31. Nagoshi, R.N.; Meagher, R.L.; Hay-Roe, M. Inferring the annual migration patterns of fall armyworm (Lepidoptera: Noctuidae) in the United States from mitochondrial haplotypes: Fall Armyworm Migration. *Ecol. Evol.* **2012**, *2*, 1458–1467. [CrossRef]
32. Nagoshi, R.N.; Silvie, P.; Meagher, R.L. Comparison of Haplotype Frequencies Differentiate Fall Armyworm (Lepidoptera: Noctuidae) Corn-Strain Populations from Florida and Brazil. *J. Econ. Entomol.* **2007**, *100*, 8. [CrossRef]
33. Troczka, B.; Zimmer, C.T.; Elias, J.; Schorn, C.; Bass, C.; Davies, T.G.E.; Field, L.M.; Williamson, M.S.; Slater, R.; Nauen, R. Resistance to diamide insecticides in diamondback moth, *Plutella xylostella* (Lepidoptera: Plutellidae) is associated with a mutation in the membrane-spanning domain of the ryanodine receptor. *Insect Biochem. Mol. Biol.* **2012**, *42*, 873–880. [CrossRef]
34. Banerjee, R.; Hasler, J.; Meagher, R.; Nagoshi, R.; Hietala, L.; Huang, F.; Narva, K.; Jurat-Fuentes, J.L. Mechanism and DNA-based detection of field-evolved resistance to transgenic Bt corn in fall armyworm (*Spodoptera frugiperda*). *Sci. Rep.* **2017**, 7. [CrossRef] [PubMed]
35. Baudron, F.; Zaman-Allah, M.A.; Chaipa, I.; Chari, N.; Chinwada, P. Understanding the factors influencing fall armyworm (*Spodoptera frugiperda* J E Smith) damage in African smallholder maize fields and quantifying its impact on yield. A case study in Eastern Zimbabwe. *Crop Prot.* **2019**, *120*, 141–150. [CrossRef]
36. Feldmann, F.; Rieckmann, U.; Winter, S. The spread of the fall armyworm *Spodoptera frugiperda* in Africa—what should be done next? *J. Plant Dis. Prot.* **2019**, *126*, 97–101. [CrossRef]
37. Gebreziher, H.G. Review on management methods of fall armyworm (*Spodoptera frugiperda* J E Smith) in Sub-Saharan Africa. *ISRN Entomol.* **2020**, *7*, 9–14.
38. Hruska, A. Fall armyworm (*Spodoptera frugiperda*) management by smallholders. *CAB Rev.* **2019**, *14*, 1–11. [CrossRef]

39. Sharanabasappa, S.; Kalleshwaraswamy, C.M.; Poorani, J.; Maruthi, M.S.; Pavithra, H.B.; Diraviam, J. Natural Enemies of *Spodoptera frugiperda* (J E Smith) (Lepidoptera: Noctuidae), a recent invasive pest on maize in South India. *Fla. Entomol.* **2019**, *102*, 619. [CrossRef]

40. Zibaee, I.; Mahmood, K.; Esmaeily, M.; Bandani, A.R.; Kristensen, M. Organophosphate and pyrethroid resistances in the tomato leaf miner *Tuta absoluta* (Lepidoptera: Gelechiidae) from Iran. *J. Appl. Entomol.* **2018**, *142*, 181–191. [CrossRef]

41. Blanco, C.; Chiaravalle, W.; Dalla-Rizza, M.; Farias, J.; García-Degano, M.; Gastaminza, G.; Mota-Sánchez, D.; Murúa, M.; Omoto, C.; Pieralisi, B.; et al. Current situation of pests targeted by Bt crops in Latin America. *Curr. Opin. Insect Sci.* **2016**, *15*, 131–138. [CrossRef]

42. Gutiérrez-Moreno, R.; Mota-Sanchez, D.; Blanco, C.A.; Whalon, M.E.; Terán-Santofimio, H.; Rodriguez-Maciel, J.C.; DiFonzo, C. Field-Evolved Resistance of the Fall Armyworm (Lepidoptera: Noctuidae) to Synthetic Insecticides in Puerto Rico and Mexico. *J. Econ. Entomol.* **2019**, *112*, 792–802. [CrossRef]

43. Bolzan, A.; Padovez, F.E.; Nascimento, A.R.; Kaiser, I.S.; Lira, E.C.; Amaral, F.S.; Kanno, R.H.; Malaquias, J.B.; Omoto, C. Selection and characterization of the inheritance of resistance of *Spodoptera frugiperda* (Lepidoptera: Noctuidae) to chlorantraniliprole and cross-resistance to other diamide insecticides. *Pest Manag. Sci.* **2019**, *75*, 2682–2689. [CrossRef]

44. Diez-Rodríguez, G.I.; Omoto, C. Herança da resistência de *Spodoptera frugiperda* (J E Smith) (Lepidoptera: Noctuidae) a lambda-cialotrina. *Neotrop. Entomol.* **2001**, *30*, 311–316. [CrossRef]

45. Do Nascimento, A.R.B.; Farias, J.R.; Bernardi, D.; Horikoshi, R.J.; Omoto, C. Genetic basis of *Spodoptera frugiperda* (Lepidoptera: Noctuidae) resistance to the chitin synthesis inhibitor lufenuron: Inheritance of lufenuron resistance in *Spodoptera frugiperda*. *Pest Manag. Sci.* **2016**, *72*, 810–815. [CrossRef] [PubMed]

46. Okuma, D.M.; Bernardi, D.; Horikoshi, R.J.; Bernardi, O.; Silva, A.P.; Omoto, C. Inheritance and fitness costs of *Spodoptera frugiperda* (Lepidoptera: Noctuidae) resistance to spinosad in Brazil. *Pest Manag. Sci.* **2017**, *74*, 1441–1448. [CrossRef] [PubMed]

47. Lira, E.C.; Bolzan, A.; Nascimento, A.R.; Amaral, F.S.; Kanno, R.H.; Kaiser, I.S.; Omoto, C. Resistance of *Spodoptera frugiperda* (Lepidoptera: Noctuidae) to spinetoram: Inheritance and cross-resistance to spinosad. *Pest Manag. Sci.* **2020**, *76*, 2674–2680. [CrossRef]

48. Nascimento, A.R.B.; Fresia, P.; Cônsoli, F.L.; Omoto, C. Comparative transcriptome analysis of lufenuron-resistant and susceptible strains of *Spodoptera frugiperda* (Lepidoptera: Noctuidae). *BMC Genom.* **2015**, *16*, 985. [CrossRef] [PubMed]

49. O'Reilly, A.O.; Khambay, B.P.S.; Williamson, M.S.; Field, L.M.; WAllace, B.A.; Davies, T.G.E. Modelling insecticide-binding sites in the voltage-gated sodium channel. *Biochem. J.* **2006**, *396*, 255–263. [CrossRef]

50. Soderlund, D.M. Molecular mechanisms of pyrethroid insecticide neurotoxicity: Recent advances. *Arch. Toxicol.* **2012**, *86*, 165–181. [CrossRef]

51. Soderlund, D.M.; Knipple, D.C. The molecular biology of knockdown resistance to pyrethroid insecticides. *Insect Biochem. Mol. Biol.* **2003**, *33*, 563–577. [CrossRef]

52. Williamson, M.S.; Martinez-Torres, D.; Hick, C.A.; Devonshire, A.L. Identification of mutations in the housefly para-type sodium channel gene associated with knockdown resistance (kdr) to pyrethroid insecticides. *Mol. Genet. Genom.* **1996**, *252*, 51–60. [CrossRef]

53. Rinkevich, F.D.; Du, Y.; Dong, K. Diversity and convergence of sodium channel mutations involved in resistance to pyrethroids. *Pestic. Biochem. Physiol.* **2013**, *106*, 93–100. [CrossRef]

54. Zhang, L.; Liu, B.; Zheng, W.; Liu, C.; Zhang, D.; Zhao, S.; Xu, P.; Wilson, K.; Withers, A.; Jones, C.M.; et al. High-depth resequencing reveals hybrid population and insecticide resistance characteristics of fall armyworm (*Spodoptera frugiperda*) invading China. *BioRxiv* **2019**, 813154. [CrossRef]

55. Harel, M.; Kryger, G.; Rosenberry, T.L.; Mallender, W.D.; Lewis, T.; Fletcher, R.J.; Guss, J.M.; Silman, I.; Sussman, J.L. Three-dimensional structures of Drosophila melanogaster acetylcholinesterase and of its complexes with two potent inhibitors. *Protein Sci.* **2000**, *9*, 1063–1072. [CrossRef] [PubMed]

56. Baek, J.H.; Kim, J.I.; Lee, D.W.; Chung, B.K.; Miyata, T.; Lee, S.H. Identification and characterization of ace1-type acetylcholinesterase likely associated with organophosphate resistance in *Plutella xylostella*. *Pestic. Biochem. Physiol.* **2005**, *81*, 164–175. [CrossRef]

57. Cassanelli, S.; Reyes, M.; Rault, M.; Carlo Manicardi, G.; Sauphanor, B. Acetylcholinesterase mutation in an insecticide-resistant population of the codling moth *Cydia pomonella (L.)*. *Insect Biochem. Mol. Biol.* **2006**, *36*, 642–653. [CrossRef] [PubMed]

58. Haddi, K.; Berger, M.; Bielza, P.; Rapisarda, C.; Williamson, M.S.; Moores, G.; Bass, C. Mutation in the ace-1 gene of the tomato leaf miner (*Tuta absoluta*) associated with organophosphates resistance. *J. Appl. Entomol.* **2017**, *141*, 612–619. [CrossRef]

59. Jiang, D.; Du, Y.; Nomura, Y.; Wang, X.; Wu, Y.; Zhorov, B.S.; Dong, K. Mutations in the transmembrane helix S6 of domain IV confer cockroach sodium channel resistance to sodium channel blocker insecticides and local anesthetics. *Insect Biochem. Mol. Biol.* **2015**, *66*, 88–95. [CrossRef]

60. Lee, D.-W.; Choi, J.Y.; Kim, W.T.; Je, Y.H.; Song, J.T.; Chung, B.K.; Boo, K.S.; Koh, Y.H. Mutations of acetylcholinesterase1 contribute to prothiofos-resistance in *Plutella xylostella (L.)*. *Biochem. Biophys. Res. Commun.* **2007**, *353*, 591–597. [CrossRef]

61. Wang, J.; Wang, B.; Xie, Y.; Sun, S.; Gu, Z.; Ma, L.; Li, F.; Zhao, Y.; Yang, B.; Shen, W.; et al. Functional study on the mutations in the silkworm (*Bombyx mori*) acetylcholinesterase type 1 gene (*Ace1*) and its recombinant proteins. *Mol. Biol. Rep.* **2014**, *41*, 429–437. [CrossRef]

62. Richardson, E.B.; Troczka, B.J.; Gutbrod, O.; Davies, T.G.E.; Nauen, R. Diamide resistance: 10 years of lessons from lepidopteran pests. *J. Pest Sci.* **2020**, *93*, 911–928. [CrossRef]

63. Nauen, R.; Steinbach, D. Resistance to Diamide Insecticides in Lepidopteran Pests. In *Advances in Insect Control and Resistance Management*; Horowitz, A.R., Ishaaya, I., Eds.; Springer: Negev, Palestine, 2016; pp. 219–240.

64. Flagel, L.; Lee, Y.W.; Wanjugi, H.; Swarup, S.; Brown, A.; Wang, J.; Kraft, E.; Greenplate, J.; Simmons, J.; Adams, N.; et al. Mutational disruption of the *Abcc2* gene in fall armyworm, *Spodoptera frugiperda*, confers resistance to the Cry1Fa and Cry1A.105 insecticidal proteins. *Sci. Rep.* **2018**, *8*, 7255. [CrossRef]

65. Guan, F.; Zhang, J.; Shen, H.; Wang, X.; Padovan, A.; Walsh, T.K.; Tay, W.T.; Gordon, K.H.J.; James, W.; Czepak, C.; et al. Whole-genome sequencing to detect mutations associated with resistance to insecticides and Bt proteins in *Spodoptera frugiperda*. *Insect Sci.* **2020**. [CrossRef] [PubMed]

66. Liu, H.; Lan, T.; Fang, D.; Gui, F.; Wang, H.; Guo, W.; Cheng, X.; Chang, Y.; He, S.; Lyu, L.; et al. Chromosome level draft genomes of the fall armyworm, *Spodoptera frugiperda* (Lepidoptera: Noctuidae), an alien invasive pest in China. *bioRxiv* **2019**, 671560. [CrossRef]

67. Nagoshi, R.N.; Koffi, D.; Agboka, K.; Tounou, K.A.; Banerjee, R.; Jurat-Fuentes, J.L.; Meagher, R.L. Comparative molecular analyses of invasive fall armyworm in Togo reveal strong similarities to populations from the eastern United States and the Greater Antilles. *PLoS ONE* **2017**, *12*, e0181982. [CrossRef] [PubMed]

68. Nagoshi, R.N.; Fleischer, S.; Meagher, R.L.; Hay-Roe, M.; Khan, A.; Murúa, M.G.; Silvie, P.; Vergara, C.; Westbrook, J. Fall armyworm migration across the Lesser Antilles and the potential for genetic exchanges between North and South American populations. *PLoS ONE* **2017**, *12*, e0171743. [CrossRef]

69. Guedes, R.N.C. Insecticide resistance, control failure likelihood and the first law of geography. *Pest Manag. Sci.* **2017**, *73*, 479–484. [CrossRef]

70. Jansson, R.K.; Brown, R.; Cartwright, B.; Cox, D.; Dunbar, D.M.; Dybas, R.A.; Eckel, C.; Lasota, J.A.; Mookerjee, P.K.; Norton, J.A.; et al. Emamectin benzoate: A novel avermectin derivative for control of lepidopterous pests. In *Proceedings of the 3rd International Workshop on Management of Diamondback Moth and Other Crucifer Pests*; MARDI: Kuala Lumpur, Malaysia, 1997; pp. 1–7.

71. Burtet, L.M.; Bernardi, O.; Melo, A.A.; Pes, M.P.; Strahl, T.T.; Guedes, J.V. Managing fall armyworm, *Spodoptera frugiperda* (Lepidoptera: Noctuidae), with Bt maize and insecticides in southern Brazil: Managing, S. frugiperda with Bt maize and insecticides. *Pest Manag. Sci.* **2017**, *73*, 2569–2577. [CrossRef]

72. Durham, E.W.; Scharf, M.E.; Siegfried, B.D. Toxicity and neurophysiological effects of fipronil and its oxidative sulfone metabolite on European corn borer larvae (Lepidoptera: Crambidae). *Pestic. Biochem. Physiol.* **2001**, *71*, 97–106. [CrossRef]

Article

A Novel Insecticidal Molecule Extracted from *Alpinia galanga* with Potential to Control the Pest Insect *Spodoptera frugiperda*

Torranis Ruttanaphan [1,2], **Georges de Sousa** [2], **Anchulee Pengsook** [3], **Wanchai Pluempanupat** [3], **Hannah-Isadora Huditz** [2], **Vasakorn Bullangpoti** [1] **and Gaëlle Le Goff** [2,*] 

[1] Animal Toxicology and Physiology Specialty Research Unit, Department of Zoology, Faculty of Science, Kasetsart University, Bangkok 10900, Thailand; torranis40@gmail.com (T.R.); fscivkb@ku.ac.th (V.B.)

[2] Université Côte d'Azur, INRAE, CNRS, ISA, F-06903 Sophia Antipolis, France; georges.de-sousa@inrae.fr (G.d.S.); hannah.huditz@hotmail.de (H.-I.H.)

[3] Department of Chemistry and Center of Excellence for Innovation in Chemistry, Faculty of Science, and Special Research Unit for Advanced Magnetic Resonance, Kasetsart University, Bangkok 10900, Thailand; a_ampere@hotmail.com (A.P.); fsciwcp@ku.ac.th (W.P.)

* Correspondence: gaelle.le-goff@inrae.fr; Tel.: +33-492-386578

Received: 9 September 2020; Accepted: 8 October 2020; Published: 11 October 2020

Simple Summary: The fall armyworm is an insect pest that feeds on many plants, including plants of agronomic importance, such as corn and rice. In addition, it has developed resistance to the main families of synthetic insecticides. There is, therefore, a need to find new, more environmentally friendly molecules to control this pest. We have extracted a molecule from greater galangal and tested its activity as an insecticide on the fall armyworm. This natural molecule causes larval growth inhibition and larval developmental abnormalities. To understand its action, a cell model with Sf9 cells was used. The molecule is much more toxic to insect cells than to human cells. It affects cell proliferation and induces cell death. This study demonstrates that a molecule extracted from an edible plant may have potential in the future development of botanical insecticides for the control of insect pests.

Abstract: *Spodoptera frugiperda*, a highly polyphagous insect pest from America, has recently invaded and widely spread throughout Africa and Asia. Effective and environmentally safe tools are needed for successful pest management of this invasive species. Natural molecules extracted from plants offer this possibility. Our study aimed to determine the insecticidal efficacy of a new molecule extracted from *Alpinia galanga* rhizome, the 1′S-1′-acetoxychavicol acetate (ACA). The toxicity of ACA was assessed by topical application on early third-instar larvae of *S. frugiperda*. Results showed that ACA caused significant larval growth inhibition and larval developmental abnormalities. In order to further explore the effects of this molecule, experiments have been performed at the cellular level using Sf9 model cells. ACA exhibited higher toxicity on Sf9 cells as compared to azadirachtin and was 38-fold less toxic on HepG2 cells. Inhibition of cell proliferation was observed at sublethal concentrations of ACA and was associated with cellular morphological changes and nuclear condensation. In addition, ACA induced caspase-3 activity. RT-qPCR experiments reveal that ACA induces the expression of several caspase genes. This first study on the effects of ACA on *S. frugiperda* larvae and cells provides evidence that ACA may have potential as a botanical insecticide for the control of *S. frugiperda*.

Keywords: 1′S-1′-Acetoxychavicol acetate; *Alpinia galanga*; *Spodoptera frugiperda*; Sf9 cells; botanical pesticide

---

1. Introduction

Spodoptera frugiperda (Smith) (Lepidoptera: Noctuidae), a major agricultural insect pest native to the American continent, has recently invaded Africa and Asia [1,2]. *S. frugiperda* causes severe yield losses in various species of economically important crops, such as maize, rice, cotton and peanut [3]. Synthetic insecticides are largely used for its management; however, their use has negative impacts on the environment and human health and led to the development of resistance. *S. frugiperda* has developed resistance against 41 different molecules (https://www.pesticideresistance.org/display.php?page=species&arId=200), resulting in pest control failures [4]. Such problems necessitate finding safer and innovative alternatives to the existing synthetic insecticides. Plants can be sources of organic molecules with insecticidal potential and advantages such as low toxicity to humans and environmental safety [5].

Plant secondary metabolites are organic compounds produced by plants and are known to play a role in adaptation to their environment [6]. They are usually subdivided according to their biosynthetic pathways in three main groups, including alkaloids, phenolics and terpenoids [6]. Plant secondary metabolites can act as insecticidal, antifeedant, repellency, oviposition deterrence and growth regulation agents against insects [5]. They have been intensively studied for the past 20 years to develop alternatives to synthetic insecticides due to their low ecological side effects [5,7].

Alpinia galanga (L.) Willd is a plant belonging to the Zingiberaceae family of Zingiberales order. Its rhizome has several active compounds, such as 1′S-1′-acetoxychavicol acetate (ACA), p-hydroxycinnamaldehyde, 1′-acetoxychavicol acetate, β-pinene, β-bisabolene and 1,8-cineole [8]. It has been used for various purposes, including antibacterial, antifungal, antiulcer, antitumor, antiallergic and antioxidant purposes. For example, it has been proposed as a new therapeutic agent in certain human cancers, such as myeloid leukemia [9], where it acts by inducing apoptosis, a natural occurring cell death process. ACA also has insecticidal activity against insect pests, such as *Bactrocera dorsalis* (Hendel), *Coptotermes gestroi* (Wasmann), *Coptotermes curvignathus* (Holmgren) and *Spodoptera litura* (Fabricius) [8,10–12]. Moreover, *A. galanga* oil could enhance the toxicity of a synthetic pyrethroid against *S. litura* [13]. In addition, *A. galanga* is a plant widespread in South East Asian countries and can be found throughout the year [14].

The aim of this study was to determine whether a compound extracted from *A. galanga*, 1′S-1′-acetoxychavicol acetate (ACA), could be used as a botanical insecticide against a major agricultural pest, *S. frugiperda*. It was a good candidate as it is a major constituent of *A. galanga* and its stereochemistry exhibited insecticidal activity [15]. The effects of this molecule and its possible mode of action were then investigated in the cell model of Sf9 cells by biochemical and molecular approaches.

2. Materials and Methods

2.1. Chemicals

Azadirachtin (95% purity), dimethyl sulfoxide (DMSO), phosphate-buffered saline (PBS), RNase A, propidium iodide (PI), penicillin-streptomycin, trypan blue solution (0.4%) and 3-(4,5-Dimethyl-2-thiazolyl)-2,5-diphenyl-2H-tetrazolium bromide (MTT) were purchased from Sigma-Aldrich, Steinheim, Germany. Alanyl-glutamine, trypsin/EDTA and fetal bovine serum were purchased from Sigma-Aldrich, St. Louis, MO, USA.

2.2. Extraction and Isolation of 1′S-1′-Acetoxychavicol Acetate

Sun-dried *A. galanga* (1.6 kg) rhizomes were powdered and sequentially extracted with hexane and ethyl acetate by soaking at room temperature for seven days. Ethyl acetate crude extract was filtered using a Buchner funnel and evaporated by a rotary evaporator. Ethyl acetate crude was fractionated by column chromatography (Kiesel gel 60 G cat No. 7731, Merck, Molsheim, France) with a gradient elution system of hexane-ethyl acetate (7:1). Eleven fractions (Fr. I–XI) were obtained. Fr. VI

(250 mg) was purified using thin layer chromatography (Silica gel 60 PF254, Merck) with hexane and ethyl acetate. 1′*S*-1′-acetoxychavicol acetate (Figure 1) was obtained and verified by spectral analysis. The analytical data of 1′*S*-1′-acetoxychavicol acetate are described as below. High resolution mass spectra were recorded using a Maxis Bruker spectrometer (Karlsruhe, Germany). ^{1}H and ^{13}C NMR spectra were recorded using a Bruker 400 MHz Advance III HD spectrometer (Karlsruhe, Germany).

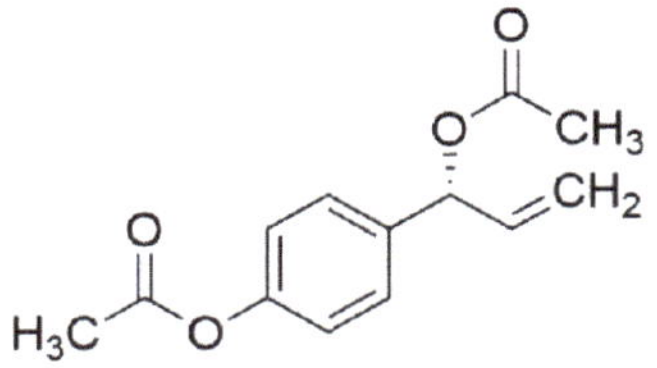

Figure 1. Chemical structure of 1′*S*-1′-acetoxychavicol acetate.

$[\alpha]_D^{20}$ − 50 (*c* = 0.5, dichloromethane). ^{1}H NMR (CDCl$_3$, 400 MHz), δ: 7.37 (2H, d, *J* = 8.7 Hz), 7.07 (2H, d, *J* = 8.7 Hz), 6.26 (1H, d, *J* = 5.7 Hz), 5.98 (1H, ddd, *J* = 17.1, 10.5, 5.8 Hz), 5.30 (1H, dt, *J* = 17.2, 1.3 Hz), 5.25 (1H, dt, *J* = 10.5, 1.2 Hz), 2.30 (3H, s), 2.11 (3H, s). ^{13}C NMR (CDCl$_3$, 100 MHz), δ: 170.0, 169.4, 150.7, 136.4, 136.3, 128.6, 121.7, 117.3, 75.8, 21.1, 21.0. HRMS (ESI) Calcd. for C$_{13}$H$_{14}$NaO$_4$ 257.0790 ([M + Na]$^+$), Found 257.0825.

2.3. Insect Rearing and Insecticide Treatments

Spodoptera frugiperda populations were provided by Dr. Emmanuelle d'Alençon (DGIMI, Université de Montpellier, INRAE, Montpellier, France). The larvae were fed ad libitum on an artificial diet [16] and maintained under controlled conditions at 24 °C and 65% relative humidity with a 16 h light: 8 h dark photoperiod as described earlier [17].

The effects of ACA on early third-instar larvae of *S. frugiperda* were determined by topical application. Serial dilutions (0.005, 0.01, 0.02 µg/µL) of ACA were prepared in acetone. Doses were applied to the dorsal thoracic region of each larva using a micropipette. Each larva received 1 µL of solution per treatment, with acetone alone as the negative control. In each experiment, 10 larvae/treatment in three replicates were used (*n* = 30 per treatment). Treated larvae were placed in a sealed plastic container with an artificial diet under controlled conditions as described above. Larval weight, abnormal development and mortality were recorded after 24, 48 and 72 h. Larvae that did not move after being prodded were considered dead. Abnormal larvae were observed as described by [18]. Relative growth rate was calculated based on the following formula described by [19] as:

Relative growth rate (%) = (Mean larval weight of treated group/Mean larval weight of control group) × 100%

2.4. Cell Culture

Sf9 cells (from Invitrogen), derived from the ovarian tissue of *S. frugiperda* pupa, were cultured in 75-cm^2 cell culture flasks (TPP, Trasadigen, Switzerland) with the Insect-XPRESS^{TM} protein-free medium (Lonza, Verviers, Belgium) and routinely maintained at 26 °C in a cell culture incubator (Memmert GmbH & Co., Schwabach, Germany). Cells were subcultured after achieving 90% confluence, and cell media were changed every 4 days.

HepG2 cells, the human hepatocellular carcinoma cell line, were provided by Dr. Georges de Sousa (INRAE, Sophia Antipolis, France). Cells were routinely maintained in William's E Medium (Sigma-Aldrich, Steinheim, Germany) supplemented with 1% alanyl-glutamine, 1% penicillin-streptomycin and 10% fetal bovine serum at 37 °C in a humidified atmosphere of 5% CO$_2$. After achieving 80–90% confluence, the cells were washed using PBS (Sigma-Aldrich,

Steinheim, Germany) and trypsinized with 0.25% Trypsin/EDTA. Cells were plated in 75-cm^2 cell culture flasks (TPP, Trasadigen, Switzerland), and cell media were changed every 3 days.

2.5. Cell Viability Assay

For cell viability assays, Sf9 cells (5×10^5 cells/mL) and HepG2 cells (2×10^5 cells/mL) were seeded onto 96-well plates (100 µL/well). Sf9 cells were treated with 0, 0.25, 0.875, 1.125, 1.75, 2.625, 3.25, 4, 5, 7.5, 10, 12.5, 15, and 17.5 µM of ACA in DMSO while HepG2 cells were treated with 5, 10, 20, 40, 80, 160, and 320 µM of ACA. After 24 h of incubation, each well received 100 µL of solution mixed with medium per treatment, with 0.25% DMSO as the negative control and azadirachtin (0–100 µM), the pure molecule, as the positive control. After 24 h of treatment, 100 µL MTT dissolved in Insect-XPRESS^{TM} medium (0.5 mg/mL) was added to each well. After an additional 2 h of incubation, the medium was discarded and DMSO was added to dissolve the formazan crystal, and the absorbance was measured at 570 and 690 nm using a microplate reader (SpectraMax, Molecular Devices). (Wokingham, UK). The relative percentage of cell viability was calculated as:

$$\text{Cell viability rate (\%)} = [(A - 0.02)/(B - 0.02)] \times 100\%$$

where A and B are the absorbances of the ACA treated cells and DMSO treated cells, respectively.

2.6. Cell Proliferation Assay

Sf9 cells (3.5×10^5 cells/mL) were seeded onto 6-well plates (2 mL/well). Serial dilutions of 0.38, 0.57, 0.88 and 1.17 µM (Inhibitory concentration (IC) IC_5, IC_{10}, IC_{20} and IC_{30}, respectively) ACA were prepared using DMSO. After 24 h of incubation, each well received 2 mL of solution mixed with medium per treatment, with 0.25% DMSO as the negative control. After 24, 48 and 72 h of treatment, cell density was evaluated using a Malassez hemocytometer (Marienfeld, Harsewinkel, Germany) counted with 0.4% Trypan blue solution staining. The morphologic characteristics of Sf9 cells were observed with an inverted phase contrast microscope (Nikon Eclipse TE2000-U, Tokyo, Japan) at 72 h after 1.17-µM ACA treatment. To examine the reversible arrest of cellular proliferation, the medium containing ACA was removed after 72 h of treatment. The cells were washed using PBS and fresh medium without ACA was added. After an additional 72 h of incubation, the cells were counted.

2.7. Caspase-3 Activity Assay

Caspase-3 activity was measured using a fluorometric activity assay kit (Molecular probe, Eugene, OR, USA) following the manufacturer's protocol. The substrate, rhodamine 110 bis-(N-CBZ-L-aspartyl-L-glutamyl-L-valyl-L-aspartic acid amide) (Z-DEVD-Rho 110), was cleaved proteolytically by caspase-3. After 24 h treatment with ACA (0.88, 1.17, and 1.75 µM corresponding to IC_{20}, IC_{30} and IC_{40}, respectively), the cells were harvested and lysed in 1 mL lysis buffer. The sample was incubated with the substrate Z-DEVD-Rho 110. Fluorescence intensity was determined using a LightCycler 480$^{®}$ (Roche Diagnostic, Inc., Mannheim, Germany) at an excitation wavelength of 490 nm and an emission wavelength of 520 nm. Values were expressed as arbitrary units of Rhodamine-110 (Rho 110) fluorescence.

2.8. Hoechst 33342 Staining

Sf9 cells (1×10^5 cells/mL) were seeded in 96-well plates, incubated for 24 h and then treated with ACA (0.88, 1.17, and 1.75 µM corresponding to IC_{20}, IC_{30} and IC_{40}, respectively) and incubated for 24 h. The cells were stained with Hoechst 33342 (Thermo Fisher Scientific Inc. Waltham, MA, USA) and incubated for 1 h at 26 °C in a cell culture incubator. The stained cells were observed using an ArrayScan™ XTI High Content Analysis (Thermo Fisher Scientific Inc. Waltham, MA, USA).

2.9. Cell Cycle Analysis

The Sf9 cell cycle analysis was carried out using fluorescence-activated cell sorting (FACS) assay (FACSCalibur, BectonDickinson, Eysins, Switzerland). The Sf9 cells were incubated for 24, 48 and 72 h with 1.17 µM of ACA (IC_{30}). Cells were resuspended and fixed with cold 80% ethanol/PBS (10 mM Na_2HPO_4, 138 mM NaCl, 2.7 mM KCl, pH 7.4). Fixed Sf9 cells were incubated in a PBS solution containing 50 µg/mL propidium iodide and 50 µg/mL RNAse A at 37 °C for 20 min. At least 10,000 cells were counted in each assay. The percentage of cells in each cell cycle phase were measured and classified in G0/G1, S and G2/M phases, depending on the intensity of the fluorescence peaks. Cells incubated with 0.25% DMSO at the same time were used as a control.

2.10. RNA Extraction, cDNA Synthesis and Real-Time Quantitative PCR (RT-qPCR)

To study gene expression of apoptosis-related genes induced by ACA, Sf9 cells were harvested after 24 h exposure to 1.17 µM ACA corresponding to IC_{30}. Total RNA was extracted using the TRIzol Reagent (Invitrogen Life Technologies, Carlsbad, CA, USA) according to the manufacturer's protocol. RNA extractions were performed in six-well plates with three biological replicates.

An amount of 1 µg total RNA was reverse transcribed using the iScript cDNA Synthesis kit (Bio-Rad). RT-qPCR reactions were performed using an Opticon monitor 2 (Bio-Rad) with the qPCR MasterMix for SYBR® Green I No ROX (Eurogentec, Seraing, Belgium). The PCR conditions were as follows: 50 °C for 2 min, 95 °C for 10 min, followed by 40 cycles of 95 °C for 30 s, 60 °C for 30 s and 72 °C for 30 s. Each reaction was performed in two technical replicates and the mean of the three independent biological replicates was calculated. All results were normalized to the glucose 6-phosphate dehydrogenase (G6PDH), ribosomal protein L4 (RpL4) and ribosomal protein L18 (L18) mRNA expression levels, identified, in one of our previous studies, as the most stably expressed reference genes under different conditions [20]. Relative expression values were calculated in R using the SATQPCR (http://satqpcr.sophia.inra.fr) described in [21]. Primer sequences and efficiencies are listed in Table 1.

Table 1. Primers used in qRT-PCR.

Name	Primer Sequence	Fragment Length (bp)	PCR Efficiencies (%)
Sf-Caspase-1-F	5′-AATGCTGGACGGAAAACAAG-3′	139	99
Sf-Caspase-1-R	5′-AACTGGCATCCTAGCGACAC-3′		
Sf-Caspase-2-F	5′-TCCAGTCCACCCTGATTTTC-3′	83	103
Sf-Caspase-2-R	5′-ACCAAGATCCACGTTTACGG-3′		
Sf-Caspase-5-F	5′-GGCCTCTACGAGTGATGGAC-3′	89	99
Sf-Caspase-5-R	5′-CGGAAGACACGTCAGTCAAA-3′		
Sf-Caspase-6-F	5′-ACCACAAGGAATGGAAGTGG-3′	149	101
Sf-Caspase-6-R	5′-GTGCTGTGTCCGGTACTTCA-3′		
Sf-Cathepsin B-F	5′-AACGGTGACTCCAAAACACC-3′	98	109
Sf-Cathepsin B-R	5′-GAGTACACGTGCTTGCCGTA-3′		
Sf-Cathepsin L-F	5′-AGTGCAGGTACAACCCCAAG-3′	146	100
Sf-Cathepsin L-R	5′-CTGGAAGGTCTCCTGTGAGG-3′		
Sf-DnaJ1-F	5′-TGAGAGAGGGAGGAGTTGGA-3′	93	95
Sf-DnaJ1-R	5′-GTCTACGACCACCGCTGAAT-3′		
Sf-p53-F	5′-GCACTTGATATCGGTGGAGAA-3′	150	96
Sf-p53-R	5′-GATCCTACAGTCACCCAGCA-3′		
Sf-TCTP-F	5′-GGACATCCTTGGCAGGTTTA-3′	147	105
Sf-TCTP-R	5′-TCCTCCTCAAGACCATGCTT-3′		
G6PD-F	5′-GGCCCTGTGGCTAACAGAAT-3′	142	98
G6PD-R	5′-CATCGTCTCTACCAAAAGGCTTC-3′		
L18-F	5′-CGTATCAACCGACCTCCACT-3′	126	108
L18-R	5′-AGGCACCTTGTAGAGCCTCA-3′		
RpL4-F	5′-CAACAAGAGGGGTTCACGAT-3′	149	98
RpL4-R	5′-GCACGATCAGTTCGGGTATC-3′		

G6PD, glucose 6-phosphate dehydrogenase; L18, ribosomal protein; RpL4, ribosomal protein.

2.11. Statistical Analysis

All experiments were conducted in three biological replications and the values are described as mean ± SE (standard error). Probit analysis was used to estimate contact toxicity as the median lethal dose (LD_{50}) values. Statistical comparisons between various groups were analyzed using Tukey's multiple range tests ($p < 0.05$ being considered significant) by the program Stat Plus v.Pro 6.5.0.0 (AnalystSoft Inc, Walnut, CA, USA).The half maximal inhibitory concentration (IC_{50}) values were calculated by log-probit analysis using the GraphPad Prism 5 program. For RT-qPCR analysis, statistical comparisons between DMSO and ACA groups were analyzed using *t*-test by the SATQPCR (http://satqpcr.sophia.inra.fr/cgi/home.cgi), described in Rancurel et al. [21].

3. Results

3.1. Extract Yield

The percentage yield of *A. galanga* for each crude extract derived from 1.6 kg of dry material was measured, and we found that ethyl acetate crude extract resulted in the highest yield percentage (1.19% yield), followed by hexane crude extract (0.43% yield). ACA (156 mg) was derived from 1 g of ethyl acetate crude extract. Ethyl acetate crude extract was brown gum, while ACA was pale yellow solid.

3.2. Effects of ACA on S. frugiperda Larvae

The effects of ACA on the growth inhibition and abnormality phenotype of third instar larvae of *S. frugiperda* were determined. Larval weights and phenotypic changes were examined after treatment with ACA and acetone (negative control). After 72 h, the larvae treated with ACA exhibited growth inhibition and abnormal phenotypes, such as a shrunk body, wrinkled body, deformed body and parts of the old cuticle remaining in the ACA-treated region (Figure 2A). The size of ACA-treated larvae was obviously smaller as compared to the larvae of the negative control. Larvae treated with 0.02 µg/µL ACA showed a significant reduction in relative growth rate of 34% ($p < 0.05$, Figure 2B) and a significant increase in abnormal phenotypes of around 27% ($p < 0.05$, Figure 2C). Additionally, the larvae treated with 0.02 µg/µL ACA weighed significantly less, with a 31.79% decrease (13.55 mg ± 1.15 vs. 19.87 mg ± 1.19, $p < 0.05$, $n = 30$).

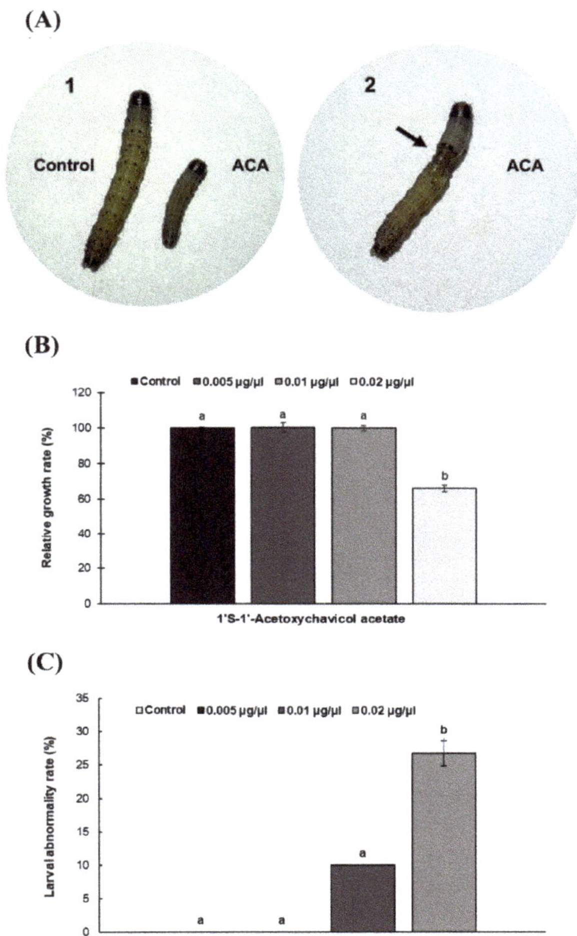

Figure 2. Effects of 1′S-1′-acetoxychavicol acetate (ACA) on *Spodoptera frugiperda* larvae; (**A**) Phenotypic changes of *S. frugiperda* larvae. (**A1**) Larval body size of *S. frugiperda* at 72 h after ACA treatment compared with acetone-treated larvae (negative control). (**A2**) Abnormalities of *S. frugiperda*. (**B**) Relative growth rate (%). (**C**) Larval abnormality rate (%). Arrow indicates larval abnormality region. Different lowercase letters indicate significantly different groups using Tukey's honestly significant difference (HSD) test ($p < 0.05$). In each experiment, 10 larvae/treatment in three replicates were used ($n = 30$ per treatment).

3.3. Cytotoxicity of ACA on Sf9 and HepG2 Cells

In order to further investigate the effect of ACA at the cellular level, we used a cellular model, Sf9, an *S. frugiperda* cell line. Cell viability was measured by MTT assay after 24 h of compound exposure. ACA exhibited a high toxicity on Sf9 cells in a dose-dependent manner (IC_{50} = 1.84 µM, Figure 3A), while azadirachtin (a commercial botanical pesticide used to control lepidopteran pests in particular), a positive control, was less toxic than ACA (percent inhibition of 100 µM azadirachtin = 11%). In order to explore the selective toxicity of ACA, experiments were performed on the non-target cellular model HepG2 cells. ACA was toxic on HepG2 cells in a dose-dependent manner (IC_{50} = 70.40 µM, Figure 3B), however this toxicity was 38-fold lower than the one observed on Sf9 cells.

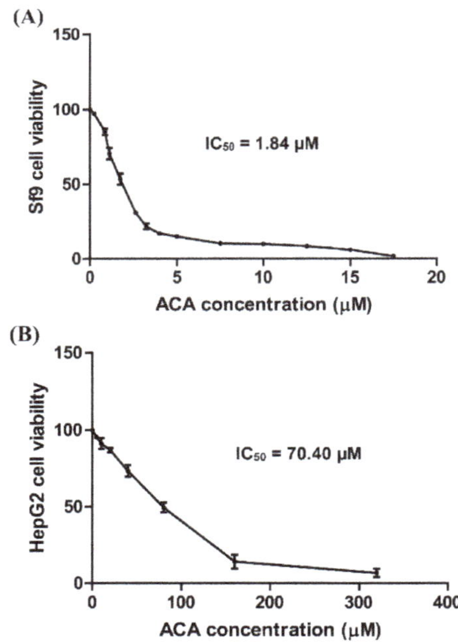

Figure 3. Effects of 1′*S*-1′-acetoxychavicol acetate on (**A**) Sf9 and (**B**) HepG2 cell viabilities at 24 h after treatment. IC$_{50}$ represents the half maximal inhibitory concentration.

3.4. Sf9 Cell Proliferation Is Inhibited by ACA

The effect of ACA on cell proliferation was examined by staining to assess the cell membrane integrity after treatment with sublethal concentrations (IC$_5$, IC$_{10}$, IC$_{20}$ and IC$_{30}$) of ACA, at various time intervals from 24 h to 72 h. The cell density in the DMSO control corresponds to the 100% proliferation rate set for each time. The proliferation rate of ACA-treated cells was calculated relative to this control. The cell density increased in the control as follows: 4.58×10^5 cells/mL, 9.75×10^5 cells/mL and 23.33×10^5 cells/mL at 24, 48, and 72 h, respectively. All treatments showed a dose-dependent decrease in the rate of cell proliferation (Figure 4A). A significant inhibition of cell proliferation was observed in the concentration range between 0.88 and 1.17 µM, with no significant difference between proliferation rate at 24 h or 72 h. Cell densities were 1.62×10^5 cells/mL, 4.83×10^5 cells/mL and 7.10×10^5 cells/mL at the concentrations of 1.17 µM after 24, 48 and 72 h of treatment, respectively. Inverted phase contrast microscopy observation revealed that Sf9 cells were irregular in shape, became larger in size and lost adhesion after 72 h treatment with 1.17 µM of ACA. In order to further confirm that ACA inhibited cell proliferation, the reversible arrest of cell proliferation was examined. The proliferation rate of the cells at 72 h after removal of ACA (1.17 µM) was significantly higher than the cells with the medium containing ACA, suggesting that the inhibitory effect of ACA on Sf9 cell proliferation is reversible (Figure 4B).

In order to evaluate the possible mechanisms involved the inhibition of cell proliferation, cell cycle monitoring and measurements of apoptosis-associated caspase activities were determined. The cell cycle was examined after 24, 48, and 72 h of 1.17 µM ACA treatments using FACS analysis. The distribution of cells in the different phases of the cell cycle was the same between control and ACA-treated cells (no significant difference) and no cell cycle arrest was observed, indicating that ACA did not affect the cell cycle (data not shown). Caspase-3 activity was evaluated in Sf9 cells after 24 h of IC$_{20}$, IC$_{30}$ and IC$_{40}$ of ACA treatments (0.88, 1.17 and 1.75 µM, respectively) using Z-DEVD-Rho 110 (Figure 4C). The results demonstrated an increase in Rho 110 fluorescence following ACA treatments, indicating that ACA induced caspase-3 activity in Sf9 cells in a dose-dependent manner. Furthermore, Hoechst 33342 staining revealed nuclear morphologic changes (Figure 4D). Nuclei of ACA-treated cells showed deeper white staining with nuclear condensation after 24 h treated with 0.88, 1.17 and 1.75 µM of ACA. The number of nuclear condensations increased in a dose-dependent manner. Altogether, these results suggest that the inhibition of cell proliferation by ACA was not due to cell cycle arrest but appeared to involve apoptosis, as evidenced by caspase activity and nuclear condensation.

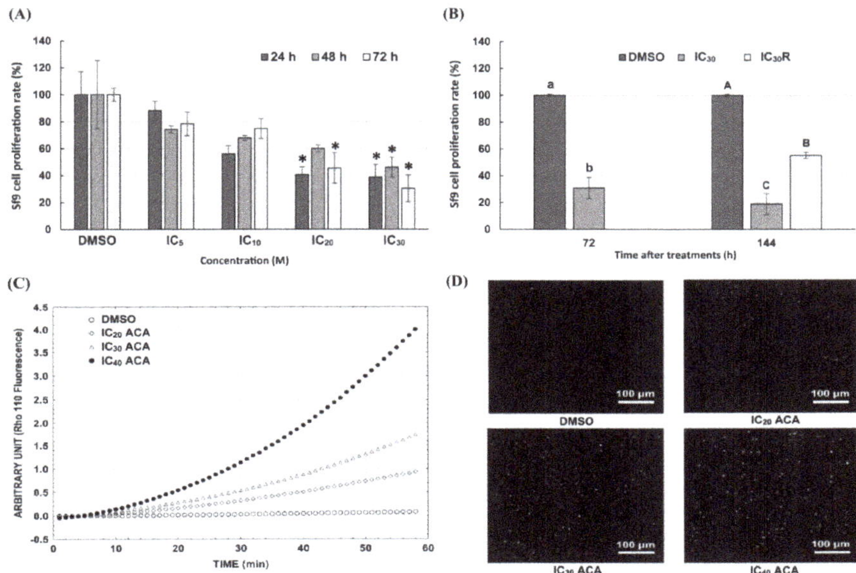

Figure 4. Effects of 1′S-1′-acetoxychavicol acetate on Sf9 cell proliferation, caspase-3 activity and nuclear condensation; (**A**) anti-proliferative effect of ACA at various concentrations and time points. (**B**) Sf9 cell proliferation was observed at 72 and 144 h after treatment by 1.17 µM of ACA. Reversible arrest of cell proliferation was observed in $IC_{30}R$ where the molecule was removed after 72 h. R stands for removal. All data were expressed as mean ± SE of three independent experiments. A *t*-test was performed to determine significance of the results as compared to the control group (0.25% dimethyl sulfoxyde DMSO). * denotes significant difference at $p < 0.05$. Tukey's honestly significant difference (HSD) test was used to compare the groups of each time intervals. Different lowercase (for 72 h) or uppercase (for 144 h) letters indicate significantly different groups ($p < 0.01$). (**C**) Caspase-3 activity of Sf9 cells after 24 h treatment with ACA. Values were expressed as arbitrary units of Rhodamine-110 (Rho 110) fluorescence. (**D**) Nuclear morphologic changes of Sf9 cells was detected by Hoechst 33342 staining.

3.5. Gene Expression Profile of Apoptosis-Related Genes after ACA Treatment of Sf9 Cells

To investigate the effect of ACA at the transcriptional level, mRNA expression levels of nine genes that play important roles in apoptosis were measured by RT-qPCR. Of these nine genes, six showed a significant differential expression after ACA treatment (Figure 5). The expression of *caspases 1, 2, 6, cathepsin B* and *p53* was significantly up-regulated compared to the DMSO-treated group following exposure to 1.17 µM ACA for 24 h (Figure 5A,B,D,E,H), while DnaJ1 was significantly downregulated (Figure 5G). These results indicate that ACA was involved in apoptosis of Sf9 cells at the transcriptional level.

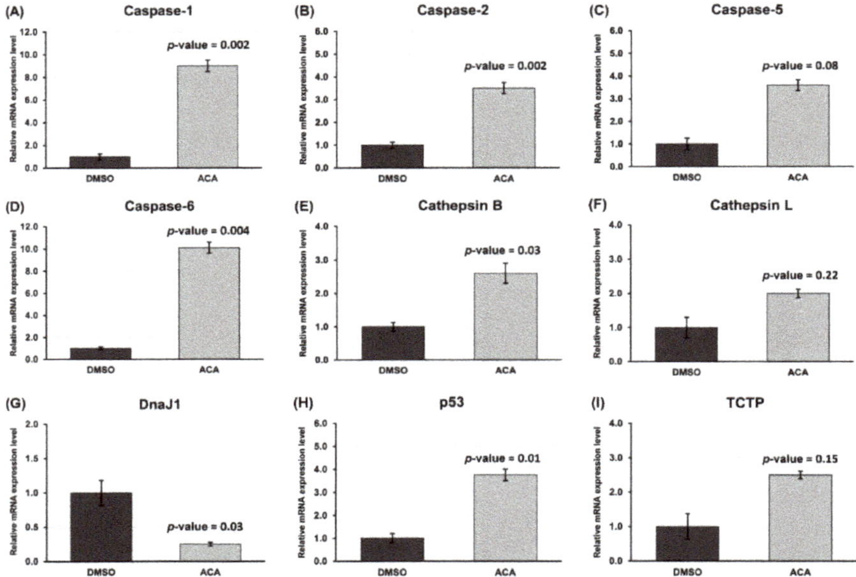

Figure 5. Effects of 1′*S*-1′-acetoxychavicol acetate on gene expression of Sf9 cells at 24 h after treatment; (**A**) Caspase-1 gene; (**B**) caspase-2; (**C**) caspase-5; (**D**) caspase-6; (**E**) cathepsin B; (**F**) cathepsin L; (**G**) DnaJ1 gene; (**H**) p53; (**I**) TCTP gene. Gene expression levels were normalized to the three reference genes; glucose 6-phosphate dehydrogenase (G6PD), ribosomal protein (L18) and ribosomal protein (RpL4). All data were expressed as mean ± SE of three independent experiments. A *t*-test was performed to determine significance of the results compared with the control group (DMSO-treated cells).

4. Discussion

Plant secondary metabolites with insecticidal properties have long been of interest as alternatives to synthetic insecticides for insect pest management [7]. Many plant secondary metabolites appear highly effective with multiple mechanisms of action while having low toxicity to non-target organisms [22,23]. So far, *A. galanga*, a plant belonging to the Zingiberaceae family, has been known for its wide range of medicinal properties [8] but little is known regarding its potential use as source of insecticide molecules. In our study, a major constituent, the 1′*S*-1′-acetoxychavicol acetate (ACA), was extracted from *A. galanga* and its insecticidal potency was evaluated on an insect pest, *S. frugiperda*.

The sustainable development of agriculture requires insecticides that are effective in insect pest management and less toxic to non-target organisms and humans [24]. ACA toxicity was evaluated in parallel on insect cells (Sf9) and mammalian cells (HepG2). HepG2 are duman hepatocellular carcinoma cells that have retained many activities found in normal hepatocytes [25,26]. Hepatocytes are increasingly being used as a cellular model for assessing the toxic potential of substances, as the liver is the principal organ associated with metabolic processes and xenobiotic toxicity [27]. Many insecticides have been tested on HepG2 cells to assess their effects on non-target organisms [28,29]. In our study, ACA was significantly less toxic to HepG2 cells than to Sf9 cells, consistent with what is expected in terms of selectivity for an insecticide molecule.

The ACA effects on the third-instar larvae of *S. frugiperda* showed an inhibition of growth and phenotypic abnormalities, such as a shrunk body, wrinkled body, deformed body and the remaining of parts of the old cuticle in the ACA-treated region. These effects are similar to those observed with insect growth disruptors, such as juvenile hormone analogues, inhibitors of chitin synthesis and ecdysone agonists. These chemicals interfere with the molecular, biochemical and physiological processes associated with insect growth and development [30–32]. However, similar mechanisms of inhibition of insect growth have also been observed in insects exposed to plant secondary metabolites. Indeed, azadirachtin is a tetranortriterpenoid limonoid isolated from *A. indica* and it could regulate growth and cocooning in *Bombyx mori* L. by inducing apoptosis in the prothoracic gland [33]. Furthermore, Shu et al. [34] reported that azadirachtin inhibited the larval growth of *S. litura* by inducing apoptosis in the midgut cells (including intestinal wall cracking, abnormal cell structure and cell death). In order to investigate the mechanisms involved in this growth inhibition, ACA effects have been evaluated on the cellular model of *S. frugiperda*, the Sf9 cells. Our results indicated that ACA had a higher toxicity than the widely used botanical insecticide azadirachtin. At sublethal concentrations, ACA caused inhibition of cell proliferation. The observed effects, including changes in cell morphology, nuclear condensation and induction of caspase-3 activity, suggested that the inhibition of cell proliferation is due to apoptosis. Similar effects of another plant compound, harmine, isolated from *Peganum harmala* were observed on Sf9 cells [35]. The apoptosis in this case had been highlighted by DNA fragmentation, cellular morphological changes, nuclear condensation, and increased caspase-3 activity [35]. Apoptosis is a naturally occurring cell death process responsible for the elimination of damaged or unwanted cells to maintain homeostasis and normal development in multicellular organisms in response to stimuli such as heat, hormones, hypoxia, radiation, virus and chemical agents [36]. Apoptosis is regulated by multiple proteins, such as caspases, inhibitors of apoptosis proteins (IAP) (including IAP, IAP-2, Survivin and Survivin-1), IAP antagonist proteins (including IBM1 and Grim-19), anti-apoptotic proteins (including DnaJ1 and TCTP), pro-apoptotic proteins (including Cathepsin B and Cathepsin L), p53 protein, etc. [37]. Among these proteins, caspases (Cysteinyl aspartate specific proteinase) form a family of cysteine proteases playing important roles in programmed cell death and can be classified, according to their biological functions, into two groups—initiator and effector caspases [38]. Once activated, initiator caspases initiate the apoptosis processes and effector caspases cleave various cellular substrates, ultimately leading to apoptosis [39]. Six caspases of lepidopteran insects have been classified based on phylogenetic analyses (caspase-1 to -6) [40]. Among them, caspase-3 is known as an effector caspase in apoptosis because of its pivotal roles, such as catalyzing the cleavage of various cellular proteins during apoptosis, coordinating the dismantling of cellular structures, chromatin condensation, DNA fragmentation, cellular morphological changes and the formation of apoptotic bodies [36]. The differential expression of caspases after exposure to chemicals or pathogen infection illustrated their various responses to external stimuli [41]. Moreover, the apoptosis, process also occurs via caspase-independent apoptotic pathways. For example, azadirachtin can induce mitochondria-mediated apoptosis by causing the opening of mitochondrial permeability transition pores and loss of mitochondrial membrane potential, ultimately leading to the release of cytochrome-c in the cytoplasm and activating caspases in the Sf9 cells [42]. The lysosomal signaling pathway also plays a role in the process of apoptosis. It was shown that azadirachtin induced lysosomal membrane permeabilization, cathepsin L releasing to cytosol and activation of caspase-3 in Sf9 cells [43]. The induced apoptosis in insect cells will help to elucidate the inhibition mechanism of cellular proliferation that could also inhibit the growth of insect pests by preventing metamorphosis [44]. Our results were in favor of caspase-dependent apoptosis with, in particular, the increased caspase-3 activity and the over-expression of p53 and several caspases at the transcriptomic level. The regulation of caspase expression has been shown to be different depending on the chemical compounds that induce apoptosis [41]. In addition, the overexpression of cathepsin B also suggested that lysosomal apoptotic pathways are activated. One gene, DnaJ1, was down-regulated after ACA exposure. DnaJ homolog subfamily A member1 (DnaJ1) belongs to the heat shock 40 kDa protein family, known to play an important role in neuroprotection and hematopoiesis of *Drosophila melanogaster* Meigen [45,46]. Recently, Shu and colleagues [47] discovered a new role for DnaJ1 as an anti-apoptotic protein. In their study, the level of DnaJ1 expression was down-regulated in Sf9 cells undergoing apoptosis following treatment with azadirachtin. When DnaJ1 is knocked out, induced apoptosis is increased. A similar decrease in the mRNA expression of DnaJ1 was observed in our study following exposure to ACA, confirming that ACA induces apoptosis in Sf9 cells.

The present study revealed that ACA from *A. galanga* exhibited an insecticidal activity against *S. frugiperda*. However, to achieve a more efficient integrated pest management, further studies are required to determine the efficacy and stability of ACA in field trails and its effects on natural enemies.

5. Conclusions

Fall armywormhas developed resistance to many insecticide molecules. Its invasiveness makes it a major pest worldwide. The invasive populations already present resistance [48], so there is a real need to find new molecules, and ACA could potentially be one of them. It responds to the renewed interest in natural molecules that are not harmful to humans and the environment. This first study on the insecticidal activity of ACA under laboratory conditions is a necessary step for a future optimization and before authorization of its usage in agriculture.

Author Contributions: Conceptualization, G.L.G., T.R., and G.d.S.; methodology, G.L.G., G.d.S., A.P., W.P., T.R., and H.-I.H.; validation, G.L.G., T.R., G.d.S., W.P., and V.B.; formal analysis, T.R.; investigation, T.R. and A.P.; resources, G.L.G., G.d.S., and A.P.; writing—original draft preparation, T.R.; writing—review and editing, G.L.G.; supervision, G.L.G.; funding acquisition, G.L.G. and V.B. All authors have read and agreed to the published version of the manuscript.

Funding: This research was supported by the Thailand Research Fund (Grant No. PHD/0043/2558), the French Government's Contribution from the Embassy of France and PHC Siam Mobility Grant (Grant No. 42852PL) and by the French Ministries of Europe and Foreign Affairs (MEAE) and of Higher Education, Research and Innovation (MESRI), as well as from the Office of the Commission for Higher Education of the Thai Ministry of Education.

Acknowledgments: The authors thank Julie Cazareth from Institut de Pharmacologie Moleculaire et Cellulaire, Sophia Antipolis, France, for technical assistance on flow cytometry and Dries Amezian for his assistance on molecular biology technique. We would like to thank Emmanuelle d'Alençon and Magali Eychenne from DGIMI, Université de Montpellier, INRAE, Montpellier, France for supplying *S. frugiperda* populations.

Conflicts of Interest: The authors declare no conflict of interest.

References

1. Chormule, A.; Shejawal, N.; Sharanabasappa; Kalleshwaraswamy, C.; Asokan, R.; Swamy, H.M. First report of the fall Armyworm, *Spodoptera frugiperda* (J. E. Smith) (Lepidoptera, Noctuidae) on sugarcane and other crops from Maharashtra, India. *J. Entomol. Zool. Stud.* **2019**, *7*, 114–117.
2. Goergen, G.; Kumar, P.L.; Sankung, S.B.; Togola, A.; Tamo, M. First Report of Outbreaks of the Fall Armyworm *Spodoptera frugiperda* (J E Smith) (Lepidoptera, Noctuidae), a New Alien Invasive Pest in West and Central Africa. *PLoS ONE* **2016**, *11*, e0165632. [CrossRef] [PubMed]
3. Murúa, M.G.; Vera, M.T.; Abraham, S.; Juarçz, M.L.; Prieto, S.; Head, G.P.; Willink, E. Fitness and Mating Compatibility of *Spodoptera frugiperda* (Lepidoptera: Noctuidae) Populations from Different Host Plant Species and Regions in Argentina. *Ann. Entomol. Soc. Am.* **2008**, *101*, 639–649. [CrossRef]
4. Mota-Sanchez, D.; Wise, J.C. *The Arthropod Pesticide Resistance Database*; Michigan State University: East Lansing, MI, USA, 2008.
5. Isman, M.B. Botanical insecticides, deterrents, and repellents in modern agriculture and an increasingly regulated world. *Annu. Rev. Entomol.* **2006**, *51*, 45–66. [CrossRef] [PubMed]
6. Bourgaud, F.; Gravot, A.; Milesi, S.; Gontier, E. Production of plant secondary metabolites: A historical perspective. *Plant Sci.* **2001**, *161*, 839–851. [CrossRef]
7. Isman, M.B. Botanical Insecticides in the Twenty-First Century-Fulfilling Their Promise? *Annu. Rev. Entomol.* **2020**, *65*, 233–249. [CrossRef]
8. Anirban, C.; Santanu, P. A Review on Phytochemical and Pharmacological Potential of *Alpinia galanga*. *Pharmacogn. J.* **2018**, *10*, 9–15. [CrossRef]
9. Ito, K.; Nakazato, T.; Murakami, A.; Yamato, K.; Miyakawa, Y.; Yamada, T.; Hozumi, N.; Ohigashi, H.; Ikeda, Y.; Kizaki, M. Induction of apoptosis in human myeloid leukemic cells by 1'-acetoxychavicol acetate through a mitochondrial- and Fas-mediated dual mechanism. *Clin. Cancer Res.* **2004**, *10*, 2120–2130. [CrossRef]

10. Abdullah, F.; Subramanian, P.; Ibrahim, H.; Abdul Malek, S.N.; Lee, G.S.; Hong, S.L. Chemical composition, antifeedant, repellent, and toxicity activities of the rhizomes of galangal, *Alpinia galanga* against Asian subterranean termites, *Coptotermes gestroi* and *Coptotermes curvignathus* (Isoptera: Rhinotermitidae). *J. Insect Sci.* **2015**, *15*, 175. [CrossRef]

11. Datta, R.; Kaur, A.; Saraf, I.; Singh, I.P.; Kaur, S. Effect of crude extracts and purified compounds of *Alpinia galanga* on nutritional physiology of a polyphagous lepidopteran pest, *Spodoptera litura* (Fabricius). *Ecotoxicol. Environ. Saf.* **2019**, *168*, 324–329. [CrossRef]

12. Sukhirun, N.; Pluempanupat, W.; Bullangpoti, V.; Koul, O. Bioefficacy of *Alpinia galanga* (Zingiberaceae) rhizome extracts, (E)-p-acetoxycinnamyl alcohol, and (E)-p-coumaryl alcohol ethyl ether against *Bactrocera dorsalis* (Diptera: Tephritidae) and the impact on detoxification enzyme activities. *J. Econ. Entomol.* **2011**, *104*, 1534–1540. [CrossRef] [PubMed]

13. Ruttanaphan, T.; Pluempanupat, W.; Aungsirisawat, C.; Boonyarit, P.; Le Goff, G.; Bullangpoti, V. Effect of Plant Essential Oils and Their Major Constituents on Cypermethrin Tolerance Associated Detoxification Enzyme Activities in *Spodoptera litura* (Lepidoptera: Noctuidae). *J. Econ. Entomol.* **2019**, *112*, 2167–2176. [CrossRef] [PubMed]

14. Poonsri, W.; Pengsook, A.; Pluempanupat, W.; Yooboon, T.; Bullangpoti, V. Evaluation of *Alpinia galanga* (Zingiberaceae) extracts and isolated *trans*-cinnamic acid on some mosquitoes larvae. *Chem. Biol. Technol. Agric.* **2019**, *6*, 17. [CrossRef]

15. Pengsook, A.; Puangsomchit, A.; Yooboon, T.; Bullangpoti, V.; Pluempanupat, W. Insecticidal activity of isolated phenylpropanoids from *Alpinia galanga* rhizomes against *Spodoptera litura*. *Nat. Prod. Res.* **2020**, 1–5. [CrossRef]

16. Poitout, S.; Bues, R. Linolenic acid requirements of lepidoptera Noctuidae Quadrifinae Plusiinae: *Chrysodeixis chalcites* Esp., *Autographa gamma* L.' *Macdunnoughia confusa* Stph., *Trichoplusia ni* Hbn. reared on artificial diets. *Ann. Nutr. Aliment.* **1974**, *28*, 173–187. [PubMed]

17. Moné, Y.; Nhim, S.; Gimenez, S.; Legeai, F.; Seninet, I.; Parrinello, H.; Negre, N.; d'Alencon, E. Characterization and expression profiling of microRNAs in response to plant feeding in two host-plant strains of the lepidopteran pest *Spodoptera frugiperda*. *BMC Genom.* **2018**, *19*, 804. [CrossRef] [PubMed]

18. Martinez, S.S.; van Emden, H.F. Growth disruption, abnormalities and mortality of *Spodoptera littoralis* (Boisduval) (Lepidoptera: Noctuidae) caused by Azadirachtin. *Neotrop. Entomol.* **2001**, *30*, 113–125. [CrossRef]

19. Akhtar, Y.; Isman, M.B. Comparative growth inhibitory and antifeedant effects of plant extracts and pure allelochemicals on four phytophagous insect species. *J. Appl. Entomol.* **2004**, *128*, 32–38. [CrossRef]

20. Giraudo, M.; Hilliou, F.; Fricaux, T.; Audant, P.; Feyereisen, R.; Le Goff, G. Cytochrome P450s from the fall armyworm (*Spodoptera frugiperda*): Responses to plant allelochemicals and pesticides. *Insect Mol. Biol.* **2015**, *24*, 115–128. [CrossRef]

21. Rancurel, C.; van Tran, T.; Elie, C.; Hilliou, F. SATQPCR: Website for statistical analysis of real-time quantitative PCR data. *Mol. Cell Probes* **2019**, *46*, 101418. [CrossRef]

22. Selin-Rani, S.; Senthil-Nathan, S.; Thanigaivel, A.; Vasantha-Srinivasan, P.; Edwin, E.S.; Ponsankar, A.; Lija-Escaline, J.; Kalaivani, K.; Abdel-Megeed, A.; Hunter, W.B.; et al. Toxicity and physiological effect of quercetin on generalist herbivore, *Spodoptera litura* Fab. and a non-target earthworm *Eisenia fetida* Savigny. *Chemosphere* **2016**, *165*, 257–267. [CrossRef] [PubMed]

23. Vieira Ribeiro, A.; de Sá Farias, E.; Almeida Santos, A.; Andrade Filomeno, C.; Barbosa dos Santos, I.; Almeida Barbosa, L.C.; Coutinho Picanço, M. Selection of an essential oil from *Corymbia* and *Eucalyptus* plants against *Ascia monuste* and its selectivity to two non-target organisms. *Crop Prot.* **2018**, *110*, 207–213. [CrossRef]

24. Casida, J.E. The greening of pesticide-environment interactions: Some personal observations. *Environ. Health Perspect.* **2012**, *120*, 487–493. [CrossRef] [PubMed]

25. Knasmuller, S.; Parzefall, W.; Sanyal, R.; Ecker, S.; Schwab, C.; Uhl, M.; Mersch-Sundermann, V.; Williamson, G.; Hietsch, G.; Langer, T.; et al. Use of metabolically competent human hepatoma cells for the detection of mutagens and antimutagens. *Mutat. Res.* **1998**, *402*, 185–202. [CrossRef]

26. Thrift, R.N.; Forte, T.M.; Cahoon, B.E.; Shore, V.G. Characterization of lipoproteins produced by the human liver cell line, Hep G2, under defined conditions. *J. Lipid. Res.* **1986**, *27*, 236–250.

27. Acosta, D.; Sorensen, E.M.; Anuforo, D.C.; Mitchell, D.B.; Ramos, K.; Santone, K.S.; Smith, M.A. An in vitro approach to the study of target organ toxicity of drugs and chemicals. *Vitr. Cell Dev. Biol.* **1985**, *21*, 495–504. [CrossRef]

28. Dehn, P.F.; Allen-Mocherie, S.; Karek, J.; Thenappan, A. Organochlorine insecticides: Impacts on human HepG2 cytochrome P4501A, 2B activities and glutathione levels. *Toxicol. Vitr.* **2005**, *19*, 261–273. [CrossRef]

29. Yun, X.; Huang, Q.; Rao, W.; Xiao, C.; Zhang, T.; Mao, Z.; Wan, Z. A comparative assessment of cytotoxicity of commonly used agricultural insecticides to human and insect cells. *Ecotoxicol. Environ. Saf.* **2017**, *137*, 179–185. [CrossRef]

30. Davis, R.E.; Kelly, T.J.; Masler, E.P.; Fescemyer, H.W.; Thyagaraja, B.S.; Borkovec, A.B. Hormonal control of vitellogenesis in the gypsy moth, *Lymantria dispar* (L.): Suppression of haemolymph vitellogenin by the juvenile hormone analogue, methoprene. *J. Insect Physiol.* **1990**, *36*, 231–238. [CrossRef]

31. Merzendorfer, H. Chitin synthesis inhibitors: Old molecules and new developments. *Insect Sci.* **2013**, *20*, 121–138. [CrossRef]

32. Nakagawa, Y.; Minakuchi, C.; Ueno, T. Inhibition of [(3)H] ponasterone a binding by ecdysone agonists in the intact Sf-9 cell line. *Steroids* **2000**, *65*, 537–542. [CrossRef]

33. Zhang, J.; Liu, H.; Sun, Z.; Xie, J.; Zhong, G.; Yi, X. Azadirachtin induced apoptosis in the prothoracic gland in *Bombyx mori* and a pronounced Ca(2+) release effect in Sf9 cells. *Int. J. Biol. Sci.* **2017**, *13*, 1532–1539. [CrossRef] [PubMed]

34. Shu, B.; Zhang, J.; Cui, G.; Sun, R.; Yi, X.; Zhong, G. Azadirachtin Affects the Growth of *Spodoptera litura* Fabricius by Inducing Apoptosis in Larval Midgut. *Front. Physiol.* **2018**, *9*, 137. [CrossRef] [PubMed]

35. Shu, B.; Zhang, J.; Jiang, Z.; Cui, G.; Veeran, S.; Zhong, G. Harmine induced apoptosis in *Spodoptera frugiperda* Sf9 cells by activating the endogenous apoptotic pathways and inhibiting DNA topoisomerase I activity. *Pestic Biochem. Physiol.* **2019**, *155*, 26–35. [CrossRef] [PubMed]

36. Elmore, S. Apoptosis: A review of programmed cell death. *Toxicol. Pathol.* **2007**, *35*, 495–516. [CrossRef] [PubMed]

37. Shu, B.; Zhang, J.; Sethuraman, V.; Cui, G.; Yi, X.; Zhong, G. Transcriptome analysis of *Spodoptera frugiperda* Sf9 cells reveals putative apoptosis-related genes and a preliminary apoptosis mechanism induced by azadirachtin. *Sci. Rep.* **2017**, *7*, 13231. [CrossRef]

38. Cooper, D.M.; Granville, D.J.; Lowenberger, C. The insect caspases. *Apoptosis* **2009**, *14*, 247–256. [CrossRef]

39. Lord, C.E.; Gunawardena, A.H. Programmed cell death in *C. elegans*, mammals and plants. *Eur. J. Cell Biol.* **2012**, *91*, 603–613. [CrossRef]

40. Courtiade, J.; Pauchet, Y.; Vogel, H.; Heckel, D.G. A comprehensive characterization of the caspase gene family in insects from the order Lepidoptera. *BMC Genom.* **2011**, *12*, 357. [CrossRef]

41. Yu, H.; Li, Z.Q.; Ou-Yang, Y.Y.; Huang, G.H. Identification of four caspase genes from *Spodoptera exigua* (Lepidoptera: Noctuidae) and their regulations toward different apoptotic stimulations. *Insect Sci.* **2019**. [CrossRef]

42. Huang, J.; Lv, C.; Hu, M.; Zhong, G. The mitochondria-mediate apoptosis of Lepidopteran cells induced by azadirachtin. *PLoS ONE* **2013**, *8*, e58499. [CrossRef] [PubMed]

43. Wang, Z.; Cheng, X.; Meng, Q.; Wang, P.; Shu, B.; Hu, Q.; Hu, M.; Zhong, G. Azadirachtin-induced apoptosis involves lysosomal membrane permeabilization and cathepsin L release in *Spodoptera frugiperda* Sf9 cells. *Int. J. Biochem. Cell Biol.* **2015**, *64*, 126–135. [CrossRef] [PubMed]

44. Shu, B.; Wang, W.; Hu, Q.; Huang, J.; Hu, M.; Zhong, G. A comprehensive study on apoptosis induction by azadirachtin in *Spodoptera frugiperda* cultured cell line Sf9. *Arch. Insect Biochem. Physiol.* **2015**, *89*, 153–168. [CrossRef] [PubMed]

45. Miller, M.; Chen, A.; Gobert, V.; Auge, B.; Beau, M.; Burlet-Schiltz, O.; Haenlin, M.; Waltzer, L. Control of RUNX-induced repression of Notch signaling by MLF and its partner DnaJ-1 during *Drosophila* hematopoiesis. *PLoS Genet.* **2017**, *13*, e1006932. [CrossRef] [PubMed]

46. Tsou, W.L.; Ouyang, M.; Hosking, R.R.; Sutton, J.R.; Blount, J.R.; Burr, A.A.; Todi, S.V. The deubiquitinase ataxin-3 requires Rad23 and DnaJ-1 for its neuroprotective role in *Drosophila melanogaster*. *Neurobiol. Dis.* **2015**, *82*, 12–21. [CrossRef]

47. Shu, B.; Jia, J.; Zhang, J.; Sethuraman, V.; Yi, X.; Zhong, G. DnaJ homolog subfamily A member1 (DnaJ1) is a newly discovered anti-apoptotic protein regulated by azadirachtin in Sf9 cells. *BMC Genom.* **2018**, *19*, 413. [CrossRef]
48. Guan, F.; Zhang, J.; Shen, H.; Wang, X.; Padovan, A.; Walsh, T.K.; Tay, W.T.; Gordon, K.H.J.; James, W.; Czepak, C.; et al. Whole-genome sequencing to detect mutations associated with resistance to insecticides and Bt proteins in *Spodoptera frugiperda*. *Insect Sci.* **2020**. [CrossRef]

Article

Development of a Species Diagnostic Molecular Tool for an Invasive Pest, *Mythimna loreyi*, Using LAMP

Hwa Yeun Nam [1], Min Kwon [1], Hyun Ju Kim [2] and Juil Kim [1,3,*

[1] Highland Agriculture Research Institute, National Institute of Crop Science, Rural Development Administration, Pyeongchang 25342, Korea; jessienam89@gmail.com (H.Y.N.); mkwon@korea.kr (M.K.)
[2] Crop foundation Division, National Institute of Crop Science, Rural Development Administration, Wanju 55365, Korea; yaehyunj@korea.kr
[3] Program of Applied Biology, Division of Bio-resource Sciences, College of Agriculture and Life Science, Kangwon National University, Chuncheon 24341, Korea
*   Correspondence: forweek@kangwon.ac.kr; Tel.: +82-33-250-6438

Received: 20 October 2020; Accepted: 17 November 2020; Published: 19 November 2020

Simple Summary: *Mythimna loreyi* is a serious pest of grain crops and reduces yields in maize plantations. *M. loreyi* is a native species in East Asia for a long time. However, this species has recently emerged as a migration (or invasive from other Asian countries) pest of some cereal crops in Korea. Little is known about its basic biology, ecology, and it is difficult to identify the morphologically similar species, *Mythimna separate*, which occur at the cornfield in the larvae stage. Species diagnosis methods for invasive pests have been developed and utilized for this reason. Currently, the molecular biology method for diagnosing *M. loreyi* species is only using the mtCO1 universal primer (LCO1490, HCO2198) and process PCR and sequencing to compare the degree of homology. However, this method requires a lot of time and effort. In this study, we developed the LAMP (loop-mediated isothermal amplification) assay for rapid, simple, and effective species diagnosis. By analyzing the mitochondrial (mt) genome, the species-specific sequence was found at the coding region of the NADH dehydrogenase subunit 5 gene. A broad range of DNA concentration was workable in LAMP assay, in which the minimum detectable DNA concentration was 100 pg. DNA releasing method was applied, which took five minutes of incubation at 95 °C without the DNA extraction process, and only some pieces of tissue from larvae and adult samples were needed. The incidence of invasive pests is gradually diversifying. Therefore, this simple and accurate LAMP assay is possibly used in the intensive field monitoring for invasive pests and integrated management of *Mythimna loreyi*.

Abstract: The *Mythimna loreyi* (Duponchel) is one of the well-known invasive noctuid pests in Africa, Australia, and many Asian countries. However, it is difficult to identify the invasive and morphologically similar species, *Mythimna separate*, which occur at the cornfield in the larvae stage. Currently, the molecular biology method for diagnosing *M. loreyi* species is only using the mtCO1 universal primer (LCO1490, HCO2198), which requires a lot of time and effort, such as DNA extraction, PCR, electrophoresis, and sequencing. In this study, the LAMP assay was developed for rapid, simple, effective species identification. By analyzing the mitochondrial genome, the species-specific sequence was found at the coding region of the NADH dehydrogenase subunit 5 gene. Based on this unique sequence, four LAMP primers and two loop primers were designed. The F3 and B3 primers were able to diagnose species-specific, in general, and multiplex PCR and specifically reacted within the inner primers in LAMP assay. The optimal incubation condition of the LAMP assay was 61 °C for 60 min with four LAMP primers, though additional loop primers, BF and LF, did not significantly shorten the amplification time. The broad range of DNA concentration was workable in LAMP assay, in which the minimum detectable DNA concentration was 100 pg. DNA releasing method was applied, which took five minutes of incubation at 95 °C without the DNA extraction process. Only some pieces of tissue of larvae and adult samples were needed to extract DNA. The incidence of invasive pests is gradually diversifying. Therefore, this simple and accurate LAMP assay is possibly applied in the intensive field monitoring for invasive pests and integrated management of *Mythimna loreyi*.

Keywords: *Mythimna loreyi*; rice armyworm; invasive pest; LAMP; diagnostic PCR

1. Introduction

The *Mythimna loreyi* (Duponchel) (often called the cosmopolitan) is a noctuid pest of grain crops found in Africa, Australia, the Near East, and the Middle East and undergoes multiple generations per year [1–4]. *M. loreyi* feeds on various host plants, including rice, wheat, maize, sugarcane, barley, sorghum, and others, which have a large effect on female fecundity. The fecundity of female moths is greatest when the larvae feed on maize in Egypt [5]. Since *M. loreyi* is facilitated to breed, some researchers focus on the identification of products secreted by the adult corpora allata [6]. Not only the physiological understanding of this species [7–10] but also ecology-based developmental characteristics [11] and flight performance [12] have been studied. For the biological control, *M. loreyi* densovirus (MIDNVs) was isolated in Egypt and characterized [13,14]. The outbreak of this pest and the damage to crops have been proliferated, particularly in some Asian countries. In Japan, *M. loreyi* typically occurs together with *M. separate* (Walker) that has significant negative effects on crop production [15]. Besides, *M. loreyi* has begun to occur and damage host plants together with *M. separate* [16].

During a couple of years in 2019 and 2020, there have been reports that cornfield has been damaged by the larvae of *M. loreyi* in Korea. This indicates that there is a possibility that *M. loreyi* can change into a sporadic pest, which can cause serious damage to crops. As per these cases, *M. loreyi* is occasionally damaged with its sister species, *M. separate*. However, it is difficult to distinguish two species at the cornfield in the larvae stage. Only one molecular diagnostic tool has been studied to distinguish *M. loreyi*, which is based on the sequencing of part of the mitochondrial COI gene [17]. However, only one mutation exists within the 658 bp amplicon. Moreover, high sequence similarity has been shown between *M. loreyi* and *M. separata*, which indicates the limitation of diagnosis of this pest.

In this study, we developed a simpler technique, termed loop-mediated isothermal amplification assay (LAMP). It is also widely used for the rapid and accurate identification of pest species [18–20]. The LAMP is a rapid, simple, effective, and specific amplification of DNA compared to real-time PCR based on the mitochondrial gene. It is performed under isothermal conditions that require a set of four primers, a strand-displacing DNA polymerase, and a water bath or heat block to maintain the temperature at about 65 °C following a one-time denaturation at 95 °C [21] or one-step incubation at about 65 °C [22].

Following the first infestation of *M. loreyi* in Korea, there is great demand from agricultural research, extension services, and farmers for diagnostic methods for these species. Therefore, we present a LAMP-based method for specimens collected in Korea and other sequences from GenBank. This method should be useful in assisting the effective pest management of *M. loreyi*.

2. Materials and Methods

2.1. Sample Collection and Mitochondrial Genome Sequencing

The larval stage of *Mythimna loreyi* Korean populations was collected from Hadong (35°02′17″ N, 127°47′12″ E) in a cornfield, 2019. Some larvae reared in the lab for morphological conformation in the adult stage, and the genomic DNA of several of the individual larva was directly extracted with DNAzol (Molecular Research Center, Cincinnati, OH, USA) and quantified by Nanodrop (NanoDrop Technologies, Wilmington, DE, USA). Besides, the genomic DNA of over 20 larvae or adults was extracted (population pooled genomic DNA). Universal primers (LCO1490 and HCO2198) were used with *M. loreyi's* individual and pooled genomic DNA as templates in 20 μL PCR reaction containing 1 U TOYOBO KOD—FX TaqTM (Toyobo Life Science, Osaka, Japan), 2X buffer (with 15 mM MgCl2), 0.2 mM each dNTP, 0.5 μM each primer, and 100 ng genomic DNA [23]. The PCR products were directly sequenced (chromatogram) to verify the nucleotide polymorphism, and no mutation was found in intraspecies. For mitochondrial genome sequencing, the Miseq platform was used, and more than 1 Gb was sequenced. To assemble these data, the CLC Assembly Cell package (version 4.2.1) was used. After trimming raw data using CLC quality trim (ver. 4.21), the assembly was accomplished using the CLC de novo assembler with dnaLCW. Assembled sequences were confirmed by BLASTZ [24]. The GeSeq program was used for annotation [25], and the result was manually checked based on the alignment of other Noctuidae species mitochondrial genomes using MEGA 7 [26].

2.2. Phylogenetic Analysis and Primer Design

Molecular phylogenetic analysis of mitochondrion genomes was inferred by using the maximum likelihood method implemented by MEGA 7 with bootstrapping [26,27]. Mitochondrial genome sequences of other Noctuidae species were downloaded from GenBank, NCBI. For comparative analysis, mitochondrial genomes were aligned using mVISTA [28,29]. Based on the global alignment result, partial sequences were re-aligned for LAMP primer design using PrimerExplorer V5.

2.3. LAMP and PCR

WarmStart® LAMP Kit (New England Biolabs, Ipswich, UK) was used for the LAMP assay. The general protocol of LAMP was performed following the manufacture's guidelines in a 25 μL reaction mixture. For the general PCR, TOYOBO KOD—FX TaqTM (Toyobo Life Science, Osaka, Japan) was used in this study. Appropriate primers were used with the following PCR amplification protocol: a 2 min denaturing step at 94 °C and a PCR amplification cycle consisting of denaturing at 94 °C for 20 s, annealing at 60 °C for 20 s, and extension at 68 °C for 30 s, which was repeated 35 times. The amplified DNA fragments were separated using 1.5% agarose gel electrophoresis and visualized with SYBR green (Life Technologies, Grand Island, NY, USA). The larval stage of *M. loreyi* samples, which were collected from Hadong (Korea) in a cornfield (2019), was used. Pheromone traps were used for adults' sample collection, such as *M. separate, Agrotis segetum, Spodoptera frugiperda, Spodoptera exigua, Spodoptera litura*, and *Helicoverpa armigera*. Traps were set in Pyeongchang (37°40'53" N, 128°43'49" E), Hongchen (37°43'35" N, 128°24'33" E), and Gangneung (37°36'56" N, 128°45'59" E) [30]. DNA samples were prepared using DNAzol (Molecular Research Center, Cincinnati, OH, USA) from trapped adults. We used universal primers (LCO1490 and HCO2198) with each DNA sample and pooled genomic DNA as templates in a 20 μL PCR reaction containing 1 U TOYOBO KOD—FX TaqTM (Toyobo Life Science, Osaka, Japan), 2X buffer (with 15 mM MgCl2), 0.2 mM each dNTP, 0.5 μM each primer, and 100 ng genomic DNA [23]. The PCR products were directly sequenced to verify the mutation of each population using RAW_F3S and RAW_B3 primer set (MACROGEN). Three biological DNA samples that were collected from the fields were used in each LAMP and PCR to validate the reliability of the LAMP condition.

3. Results

3.1. Mitochondrial Genome Sequencing and Primer Design

A 15,320 bp of mitochondrial genome verified after trimming from about 2.2 Gb (7,312,504 reads) nucleotide sequences information was obtained through *Miseq*. The mitochondrial genome *of Mythimna loreyi* was assembled (GenBank MT506351). The mitochondrial genome included 13 protein-coding genes: NADH dehydrogenase components (complex *I*, ND), cytochrome oxidase subunits (complex VI, COX), cytochrome oxidase b (CYPB) and two ATP synthases, and two ribosomal RNA genes and 22 transfer RNAs (Supplementary Figure S1).

As a result of MegaBLAST, the most homologous species was an allied species, *Mythimna separate*, which showed 93.7% similarity. The genus of Spodoptera [31], such as *Spodoptera exigua*, *S. litura*, *S. frugiperda*, which are possibly found together with *M. loreyi* in the cornfields, showed about 89 to 90% homology based on the mitochondrial genome sequence. *Agrotis segetum* [32], which, in particular, occurs and damage at the corn seedling stage, showed 90.6% similarity to *M. loreyi* based on mitochondrial genome sequence (data not shown).

The phylogenetic relationship between 15 mitochondrial genomes of 14 species was examined (Figure 1A) to verify a specific nucleotide sequence that only *M. loreyi* possessed among the related species with high gene similarity or a morphologically similar pest. The result of the phylogenetic relationship was mostly similar to the megaBLAST result. Based on the mVISTA alignment results, conserved regions among Noctuidae species and variable regions were observed (Figure 1B). By combining the two results, the partial sequence in eight species of nine ND5 mitochondrial genome was re-aligned to design the specific primer of the *M. loreyi*. Finally, four essential primers and two loop primers were designed (Figure 2 and Table 1).

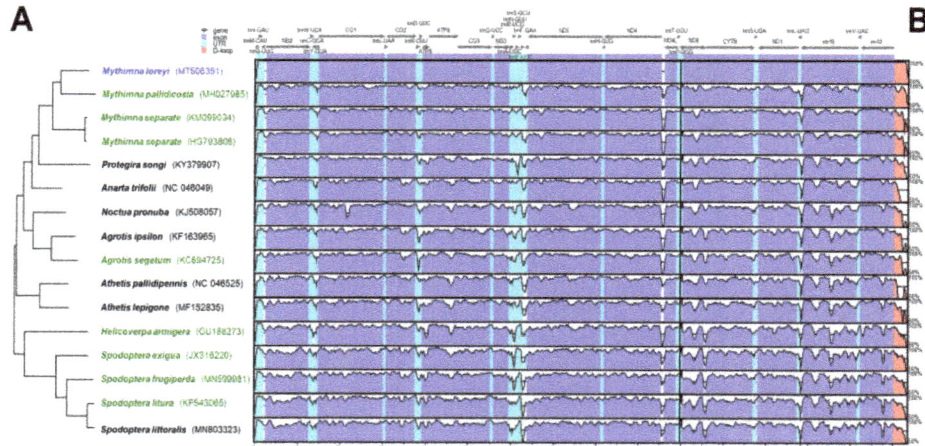

Figure 1. Comparison of entire mitochondrial genomes of some Noctuidae pests, including newly sequenced *Mythimna loreyi*. (**A**) Phylogenetic relationship inferred using maximum likelihood under MEGA7. (**B**) Schematic diagram of the genes and their flanking regions, showing the sequence diversity in mVISTA. UTR, D-loop denotes untranslated region and displacement-loop, respectively. Eight green colored mitochondrial genome sequences were re-aligned for primer design with that of the target species, *M. loreyi* (Figure 2).

Figure 2. Location of primers and primer binding regions on partial sequence of some Noctuidae pests' mtDNA for species identification of *M. loreyi*. Inner primer, FIP, consists of F1c (complementary sequences of F1) and F2. Another inner primer, BIP, is also composed of B1 and B2c (complementary sequences of B2). Essential four LAMP primers (F3, FIP, BIP, and B3) generate the dumbbell structure, and two loop primers, LF and LB, possibly accelerate the LAMP (loop-mediated isothermal amplification) reaction. Used primer information is documented in Table 1.

Table 1. Primer list for LAMP (loop-mediated isothermal amplification) and PCR in this study.

Purpose	Primers	Sequence (5′–> 3′)
for LAMP		
	RAW_F3	GTAAATTTATTAACAGAATAAATCCCC [1]
	RAW_B3	CTTCTACTTTAGTAACTGCGGGA
	RAW_FIP	TGGGGTAATTATTCATATAATAAATGATAATATATAATCTAATTCCCCCTATAAAACG
	RAW_BIP	TAGCATGAGTTAATAAATGAAAAAAAAGGTTTAATAATAAGAATTTTAAGAATGGGT
	RAW_LF	CATATAATAAATGATATACAAGATATT
	RAW_LB	AATGAAAAAAAGCTAAATCAGGTAA
for PCR		
	Spo_ace1UF2	AGGATGAAGAGAAATTTATAGAGGAT
	Spo_ace1UR1	TCACCAAACACTGTATCTATAATTTG
	LCO1490	GGTCAACAAATCATAAAGATATTGG
	HCO2198	TAAACTTCAGGCTGACCAAAAAATCA
	RAW_F3S	CCCAAACCCTCTATATAATTCTCT

[1] 'G's at the 5′ end depicted in red were added to adjust the primer melting temperature.

Among the six primers, F3 is the specific primer that enables the diagnosis of *M. loreyi*, and only *M. loreyi* had the CCCC sequences in four priming regions, which are marked in the red box. A total of four populations that were collected from different regions had the same nucleotide sequence. Therefore, it could be sufficiently used for species diagnosis. The basic species diagnosis primer production strategy is the same as the previously reported species diagnosis development method of *S. frugiperda* [33]. We targeted the open reading frame (ORF) region because the ORF region is more intraspecies conserved than the non-coding region.

3.2. Diagnostic LAMP and PCR

As previously reported, the sensitivity of LAMP may vary depending on the temperature and reaction time [33]. Therefore, the reaction was performed at 65, 63, and 61 °C to find the optimal reaction conditions (Figure 3).

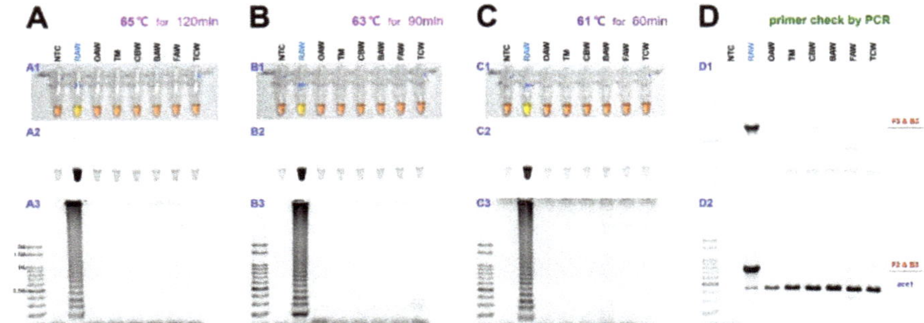

Figure 3. The sensitivity of the LAMP assay results in three temperature conditions, such as 65, 63, and 61 °C, for *M. loreyi* species detected under (**A1**, **B1**, and **C1**) visible light, (**A2**, **B2**, and **C2**) ultraviolet light with SYBR Green, and (**A3**, **B3**, and **C3**) gel electrophoresis. The original pink color of the reaction mixture turned yellow in a positive reaction when the product was formed but remained pink in negative reactions. (**D**) Conventional and multiplex PCR to distinguish *M. loreyi*. The 794 bp amplicon was amplified only in *M. loreyi*, and the conserved partial sequence of ace1 type acetylcholinesterase gene was targeted as an internal reference. Abbreviations are NTC (non-template control), RAW (rice armyworm *Mythimna loreyi*), OAW (oriental armyworm *Mythimna separata*), TM (Turnip moth *Agrotis segetum*), CBW (cotton bollworm *Helicoverpa armigera*), BAW (beet armyworm *Spodoptera exigua*), FAW (fall armyworm *Spodoptera frugiperda*), and TCW (tobacco cutworm *Spodoptera litura*).

As the amount of template DNA was quantified as 50 ng and reacted at each temperature with 25 μL reaction volume, the diagnosis result was confirmed in 120 min at 65 °C and 90 min at 63 °C. The diagnostic level of reaction did not occur when the reaction performed less than the corresponding time in each temperature. Despite the relatively low temperature, at 61 °C, the diagnostic level of reaction was confirmed in only 60 min, and the false-positive reaction did not appear only once in the results of more than three repetition tests. Despite the relatively low temperature, at 61 °C, the diagnostic level of reaction was confirmed in only 60 min, and the false-positive reaction did not appear only once in three repetition tests (Figure 3C). Under 100 bp sized bands were not the LAMP products and were aggregated primers and generated even in the negative control.

In the LAMP assay, to confirm the diagnosis primer F3, it was reacted with reverse primer B3 through PCR, and the 794 bp of PCR product was identified (Figure 3(D1)). Besides, a species-specific reaction was confirmed, as we performed the multiplex PCR with a universal positive control primer set that targets ace1-type acetylcholinesterase, which can produce a positive reaction in all samples (Figure 3(D2)). The two-loop primers were tested for possible enhancement of the LAMP reaction, as suggested by Nagamine et al. [22].

As a result of reacting two-loop primers with each or two together at 61 °C for 60 min, there was no difference between the reaction result of using only four primers and shortening the reaction time (Figure 4). Even when 10 μL of the reaction solution was verified by electrophoresis after the LAMP reaction, there was no difference in band intensity. It was possible to reliably diagnose up to 100 pg when reacting using four LAMP primers without adding a loop primer (Figure 5A). Therefore, *M. loreyi* species diagnosis LAMP method that we developed could diagnose under various DNA concentration conditions from 100 ng to 100 pg. To increase the usability in the field, we cut a part of the tissue of the adult antenna or leg and put it in 30 μL distilled water, which reacts at 95 °C for 5 min, securing the template DNA without a separate DNA extraction process (Figure 5(B4)). As we measured each sample of DNA concentration obtained through this DNA releasing method, each sample showed various measurements. The species-specific diagnosis was possible in the same method as using the template DNA, which was obtained through separate DNA extraction as a positive control since it was in the range of LAMP reaction (Figure 5B).

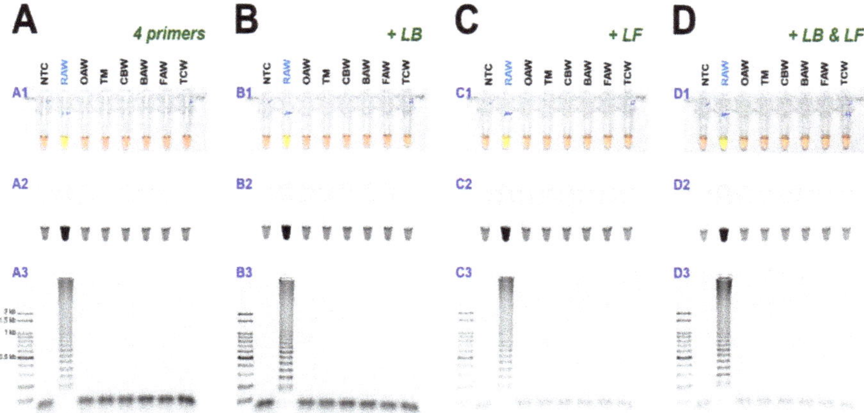

Figure 4. The LAMP assay results with (**A**) 4 primers and additional loop primers, (**B**) loop forward, LF, or (**C**) loop backward, LB, or (**D**) two-loop primers, LF and LB under (**A1**, **B1**, **C1**, and **D1**) visible light, (**A2**, **B2**, **C2**, and **D2**) ultraviolet light with SYBR Green, and (**A3**, **B3**, **C3**, and **D3**) gel electrophoresis. Abbreviations are NTC (non-template control), RAW (rice armyworm *Mythimna loreyi*), OAW (oriental armyworm *Mythimna separata*), TM (Turnip moth *Agrotis segetum*), CBW (cotton bollworm *Helicoverpa armigera*), BAW (beet armyworm *Spodoptera exigua*), FAW (fall armyworm *Spodoptera frugiperda*), and TCW (tobacco cutworm *Spodoptera litura*).

4. Discussion

Invasive pests, such as *S. frugiperda*, are increasing worldwide due to global warming and climate change [34,35]. Currently, *M. loreyi* is a serious grain pest in Africa, Australia, the Near East, and the Middle East and undergoes multiple generations per year [1–4]. Besides, *M. loreyi* is a species that originates from China and results in damage upon spreading gradually. In Korea, *M. loreyi* has been reported for a long time and maybe a native pest to Korea. Recently, there are several reports that the larvae of *M. loreyi* have damaged the cornfield in Korea. There is a possibility that *M. loreyi* can change into a sporadic pest, which can cause serious damage to crops in Korea.

As with other invasive pests found in the field, they are often similar in morphology to allied species, making the investigation of initial density and pest management difficult. Species diagnosis methods for invasive pests, such as *S. frugiperda*, have been developed and utilized for this reason [33]. Currently, the molecular biology method for diagnosing *M. loreyi* species is only using the mtCO1 universal primer (LCO1490, HCO2198) and process PCR and sequencing to compare the degree of homology. However, this method requires a lot of time and effort, such as DNA extraction, PCR, electrophoresis, and sequencing. Therefore, we developed a molecular diagnosis method that could diagnose species within a short time without a separate DNA extraction process, and only a heat block is needed that can control temperature (Figure 5). The basis diagnosis strategy is very similar to the *S. frugiperda* species diagnosis method [33]. This method can complete all experimental procedures and verify the results within 1 h and 30 min right after obtaining a sample. In this study, the results only specified adult samples, but it is possible to use larvae.

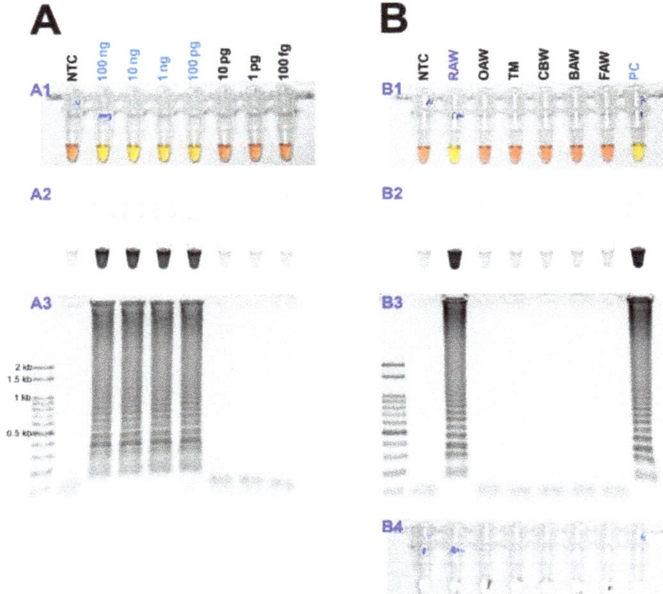

Figure 5. (**A**) Identification of the detection limit of genomic DNA in the LAMP assay from 100 ng to 100 fg under (**A1** and **B1**) visible light, (**A2** and **B2**) ultraviolet light with SYBR Green, and (**A3** and **B3**) gel electrophoresis. (**B**) The sensitivity of the LAMP assay results with the DNA releasing technique from insect tissue. Around 10 mg of the adult leg (or antenna) was incubated at 95 °C for 5 min (**B4**). Abbreviations are NTC (non-template control), RAW (rice armyworm *Mythimna loreyi*), OAW (oriental armyworm *Mythimna separata*), TM (Turnip moth *Agrotis segetum*), CBW (cotton bollworm *Helicoverpa armigera*), BAW (beet armyworm *Spodoptera exigua*), FAW (fall armyworm *Spodoptera frugiperda*), and PC (positive control, isolated DNA from *M. loreyi*).

The simplicity, accuracy, and adaptability for high throughput of the LAMP assay are distinct advantages [21,36,37]. Moreover, recently LAMP is utilized in various fields, such as many ecology studies, medical aspects, outside of the lab, and can be applied to diagnose plant viruses in insect body and insecticide-resistant gene mutation [38,39]. Besides, the diagnostic primer used for LAMP can be used for various diagnostic methods because it was possible to apply in general PCR and multiplex PCR (Figure 3). Therefore, it is feasible to diagnose a larger sample with positive control in the form of multiplex PCR, which is sufficiently modified and used in the laboratory. There are advantages and disadvantages to find a species diagnostic marker in the mitochondrial genome as well as in part of the genomic DNA. First of all, the disadvantage is that LAMP primer production is limited because there are many parts of AT-rich. It is complicated to design a primer that requires a minimum of four primers, and primers FIP and BIP are high-performance liquid chromatography (HPLC) purified primers. As an advantage, LAMP has higher amplification and efficiency and sensitivity compared to real-time PCR, and the results can be visually monitored by the naked eye either through turbidity or color change by fluorescent intercalating dye (Syber Green I) [40]. The number of copies is large, and it is possible to diagnose with a small amount of DNA or DNA releasing method, whose amplification can be accomplished with heating block. On the contrary, a species diagnosis marker in the genomic DNA should be found in exon rather than intron because it has a large variation. But it is difficult to develop marker in exon because it is often quite conserved within an allied species. Besides, the DNA releasing method can be used, but the efficiency is low when the copies of the gene are small [41].

5. Conclusions

In this study, a species diagnosis marker was designed within the mitochondrial genome to combine it with a DNA releasing method that is highly applicable to diagnose species in the field. Moreover, a significantly efficient method was developed, which targeted the ace1 gene as a positive control of the species to compare. This simple and accurate diagnosis using LAMP assay could be possibly applied in the intensive field to monitor and for the pest management of *M. loreyi*.

Supplementary Materials: The following are available online at http://www.mdpi.com/2075-4450/11/11/817/s1, Figure S1: Organization of the mitochondrial genome of *Mythimna loreyi* from Korea (GenBank MT506351). ND: NADH dehydrogenase components (Complex I) in yellow. COX: cytochrome oxidase subunits (Complex VI) in pink. ATP synthase in green. CYPB: cytochrome oxidase b in purple. Ribosomal RNA genes in red, tRNA genes in blue. Noncoding regions are not colored.

Author Contributions: Conceptualization, H.Y.N. and J.K.; methodology, H.Y.N. and J.K.; software, J.K.; validation, J.K.; formal analysis, J.K.; investigation, H.Y.N. and J.K.; resources, H.Y.N., M.K., H.J.K., and J.K.; data curation, H.Y.N. and J.K.; writing—Original draft preparation, H.Y.N. and J.K.; writing—Review and editing, H.Y.N. and J.K.; visualization, H.Y.N. and J.K.; supervision, J.K. All authors have read and agreed to the published version of the manuscript.

Funding: The Cooperative Research Program supported this study for Agriculture Science and Technology Development (Project No. PJ01509301), the Rural Development Administration, Republic of Korea.

Conflicts of Interest: The authors declare no conflict of interest.

References

1. Calora, F.B. Revision of the species *Leucania*-complex occurring in the Philippines (Lepidoptera, Noctuidae, Hadeninae). *Philipp. Agric.* **1966**, *50*, 633–723.
2. Chandler, K.J.; Benson, A.J. Evaluation of armyworm infestation in North Queensland surgarcane on crops. *Proc. Aust. Soc. Sug. Cane Technol.* **1991**, *13*, 79–82.
3. Edwards, E.D. A second sugarcane armyworm (*Leucania loreyi* (Duponchel) from Australia and the identity of *L. loreyimima* Rungs (Lepidoptera: Noctuidae). *J. Aust. Entomol. Soc.* **1992**, *31*, 105–108. [CrossRef]
4. Ganesha, S.; Rajabale, S. The *Mythimna* spp. (Lepidoptera: Noctuidae) complex on sugarcane in Mauritius. *Proc. S. Afr. Sug. Technol. Ass.* **1996**, *70*, 15–17.
5. El-Sherif, S.I. On the biology of *Leucania loreyi*, dup. (Lepidoptera, Noctuidae). *J. Appl. Entomol.* **1972**, *71*, 104–111. [CrossRef]
6. Ho, H.Y.; Tu, M.P.; Chang, C.Y.; Yin, C.M.; Kou, R. Identification of in vitro release products of corpora allata in female and male loreyi leafworms, *Leucania loreyi*. *Experientia* **1995**, *51*, 601–605. [CrossRef]
7. Hsieh, Y.C.; Hsu, E.L.; Chow, Y.S.; Kou, R. Effects of calcium channel antagonists on the corpora allata of adult male loreyi leafworm *Mythimna loreyi*: Juvenile hormone acids release and intracellular calcium level. *Arch. Insect Biochem. Physiol.* **2001**, *48*, 89–99. [CrossRef]
8. Hsieh, Y.C.; Yang, E.C.; Hsu, E.L.; Chow, Y.S.; Kou, R. Voltage-dependent calcium channels in the corpora allata of the adult male loreyi leafworm, *Mythimna loreyi*. *Insect Biochem. Mol. Biol.* **2002**, *32*, 547–557. [CrossRef]
9. Kou, R. Cholinergic regulation of the corpora allata in adult male loreyi leafworm *Mythimna loreyi*. *Arch. Insect Biochem. Physiol.* **2002**, *49*, 215–224. [CrossRef]
10. Kou, R.; Chen, S.J. Allatotropic activity in the suboesophageal ganglia and corpora cardiaca of the adult male loreyi leafworm, *Mythimna loreyi*. *Arch. Insect Biochem. Physiol.* **2000**, *43*, 78–86. [CrossRef]
11. Qin, J.; Zhang, L.; Liu, Y.; Sappington, T.W.; Cheng, Y.; Luo, L.; Jiang, X. Population Projection and Development of the *Mythimna loreyi* (Lepidoptera: Noctuidae) as Affected by Temperature: Application of an Age-Stage, Two-Sex Life Table. *J. Econ. Entomol.* **2017**, *110*, 1583–1591. [CrossRef] [PubMed]
12. Qin, J.; Liu, Y.; Zhang, L.; Cheng, Y.; Sappington, T.W.; Jiang, X. Effects of Moth Age and Rearing Temperature on the Flight Performance of the Loreyi Leafworm, *Mythimna loreyi* (Lepidoptera: Noctuidae), in Tethered and Free Flight. *J. Econ. Entomol.* **2018**, *111*, 1243–1248. [CrossRef] [PubMed]
13. El-Far, M.; Li, Y.; Fediere, G.; Abol-Ela, S.; Tijssen, P. Lack of infection of vertebrate cells by the densovirus from the maize worm *Mythimna loreyi* (MlDNV). *Virus Res.* **2014**, *99*, 17–24. [CrossRef] [PubMed]

14. Fediere, G.; El-Far, M.; Li, Y.; Bergoin, M.; Tijssen, P. Expression strategy of densonucleosis virus from *Mythimna loreyi*. *Virology* **2004**, *320*, 181–189. [CrossRef]

15. Hirai, K. The influence of rearing temperature and density on the development of two Leucania species, *M. loreyi* dup. and *L. separata* walker (Lepidoptera: Noctuidae). *Appl. Ent. Zool.* **1975**, *10*, 234–237. [CrossRef]

16. Guo, S.J.; Li, S.M.; Ma, L.P.; Zhuo, X.N. Research about biological characteristics and damage laws of *Leucania loreyi*. *J. Henan Agric. Sci.* **2003**, *9*, 37–39.

17. Jindal, V. DNA barcode reveals occurrence of *Mythimna loreyi* (Duponchel) in Punjab, India. *Indian J. Biotechnol.* **2019**, *18*, 81–84.

18. Blaser, S.; Diem, H.; von Felten, A.; Gueuning, M.; Andreou, M.; Boonham, N.; Tomlinson, J.; Müller, P.; Utzinger, J.; Frey, J.E.; et al. From laboratory to point of entry: Development and implementation of a loop-mediated isothermal amplification (LAMP)-based genetic identification system to prevent introduction of quarantine insect species. *Pest Manag. Sci.* **2018**, *74*, 1504–1512. [CrossRef]

19. Hsieh, C.H.; Wang, H.Y.; Chen, Y.F.; Ko, C.C. Loop-mediated isothermal amplification for rapid identification of biotypes B and Q of the globally invasive pest *Bemisia tabaci*, and studying population dynamics. *Pest. Manag. Sci.* **2012**, *68*, 1206–1213. [CrossRef]

20. Kim, Y.H.; Hur, J.H.; Lee, G.S.; Choi, M.Y.; Koh, Y.H. Rapid and highly accurate detection of *Drosophila suzukii*, spotted wing Drosophila (Diptera: Drosophilidae) by loop-mediated isothermal amplification assays. *J. Asia Pac. Entomol.* **2016**, *19*, 1211–1216. [CrossRef]

21. Notomi, T.; Mori, Y.; Tomita, N.; Kanda, H. Loop-mediated isothermal amplification (LAMP): Principle, features, and future prospects. *J. Microbiol.* **2015**, *53*, 1–5. [CrossRef] [PubMed]

22. Nagamine, K.; Hase, T.; Notomi, T. Accelerated reaction by loop-mediated isothermal amplification using loop primers. *Mol. Cell. Probes* **2002**, *16*, 223–229. [CrossRef] [PubMed]

23. Folmer, O.; Black, M.; Hoeh, W.; Lutz, R.; Vrijenhoek, R. DNA primers for amplification of mitochondrial cytochrome c oxidase subunit I from diverse metazoan invertebrates. *Mol. Mar. Biol. Biotechnol.* **1994**, *3*, 294–299. [PubMed]

24. Schwartz, S.; Kent, W.J.; Smit, A.; Zhang, Z.; Baertsch, R.; Hardison, R.C.; Haussler, D.; Miller, W. Human-mouse alignments with BLASTZ. *Genome Res.* **2003**, *13*, 103–107. [CrossRef]

25. Tillich, M.; Lehwark, P.; Pellizzer, T.; Ulbricht-Jones, E.S.; Fischer, A.; Bock, R.; Greiner, S. GeSeq—versatile and accurate annotation of organelle genomes. *Nucleic Acids Res.* **2017**, *45*, W6–W11. [CrossRef]

26. Kumar, S.; Stecher, G.; Tamura, K. MEGA7: Molecular Evolutionary Genetics Analysis Version 7.0 for Bigger Datasets. *Mol. Biol. Evol.* **2016**, *33*, 1870–1874. [CrossRef]

27. Sanderson, M.J.; Wojciechowski, M.F. Improved bootstrap confidence limits in large-scale phylogenies, with an example from Neo-Astragalus (Leguminosae). *Syst. Biol.* **2000**, *49*, 671–685. [CrossRef]

28. Frazer, K.A.; Pachter, L.; Poliakov, A.; Rubin, E.M.; Dubchak, I. VISTA: Computational tools for comparative genomics. *Nucleic Acids Res.* **2004**, *32*, W273–W279. [CrossRef]

29. Mayor, C.; Brudno, M.; Schwartz, J.R.; Poliakov, A.; Rubin, E.M.; Frazer, K.A.; Patcher, L.S.; Dubchak, I. VISTA: Visualizing global DNA sequence alignments of arbitrary length. *Bioinformatics* **2000**, *16*, 1046–1047. [CrossRef]

30. Kim, J.; Kwon, M.; Park, K.J.; Maharjanm, R. Monitoring of four major lepidopteran pests in Korean cornfields and management of *Helicoverpa armigera*. *Entomol. Res.* **2018**, *48*, 308–316. [CrossRef]

31. CABI Invasive Species Compendium. Available online: https://www.cabiorg/isc/search/index (accessed on 14 August 2020).

32. Erasmus, A.; Van Rensburg, J.B.J.; Van den Berg, J. Effects of Bt maize on *Agrotis segetum* (Lepidoptera: Noctuidae): A pest of maize seedlings. *Environ. Entomol.* **2010**, *39*, 702–706. [CrossRef] [PubMed]

33. Kim, J.; Nam, H.Y.; Kwon, M.; Kim, H.; Yi, H.J.; Hänniger, S.; Unbehend, M.; Heckel, D.G. Development of a simple and accurate molecular tool for *Spodoptera frugiperda* species identification using LAMP. *bioRxiv* **2020**. [CrossRef]

34. Li, X.J.; Wu, M.-F.; Ma, J.; Gao, B.-Y.; Wu, Q.-L.; Chen, A.-D.; Liu, J.; Jiang, Y.-Y.; Zhai, B.-P.; Early, R.; et al. Prediction of migratory routes of the invasive fall armyworm in eastern China using a trajectory analytical approach. *Pest Manag. Sci.* **2020**, *76*, 454–463. [CrossRef] [PubMed]

35. Goergen, G.; Kumar, P.L.; Sankung, S.B.; Togola, A.; Tamo, M. First Report of Outbreaks of the Fall Armyworm *Spodoptera frugiperda* (J E Smith) (Lepidoptera, Noctuidae), a New Alien Invasive Pest in West and Central Africa. *PLoS ONE* **2010**, *11*, e0165632. [CrossRef]

36. Mori, Y.; Notomi, T. Loop-mediated isothermal amplification (LAMP): A rapid, accurate, and costeffective diagnostic method for infectious diseases. *J. Infect. Chemother.* **2009**, *15*, 62–69. [CrossRef]
37. Zhang, X.Z.; Lowe, S.B.; Gooding, J.J. Brief review of monitoring methods for loop-mediated isothermal amplification (LAMP). *Biosens. Bioelectron.* **2014**, *61*, 491–499. [CrossRef]
38. Lee, P.L. DNA amplification in the field: Move over PCR, here comes LAMP. *Mol. Ecol. Resour.* **2017**, *17*, 138–141. [CrossRef]
39. Choi, B.H.; Hur, J.H.; Heckel, D.G.; Kim, J.; Koh, Y.H. Development of a highly accurate and sensitive diagnostic tool for pyrethroid-resistant chimeric P450 CYP337B3 of *Helicoverpa armigera* using loop-mediated isothermal amplification. *Arch. Insect Biochem. Physiol.* **2018**, *99*, e21504. [CrossRef]
40. Parida, M.; Sannarangaiah, S.; Dash, P.K.; Rao, P.V.L.; Morita, K. Loop mediated isothermal amplification (LAMP): A new generation of innovative gene amplification technique; perspectives in clinical diagnosis of infectious diseases. *Rev. Med. Virol.* **2008**, *18*, 407–421. [CrossRef]
41. Kim, J.; Cha, D.J.; Kwon, M.; Maharjan, R. Potato virus Y (PVY) detection in a single aphid by one-step RT-PCR with boiling technique. *Entomol. Res.* **2016**, *46*, 278–285. [CrossRef]

Publisher's Note: MDPI stays neutral with regard to jurisdictional claims in published maps and institutional affiliations.

Article

Screening of *Helicoverpa armigera* Mobilome Revealed Transposable Element Insertions in Insecticide Resistance Genes

Khouloud KLAI [1,2], Benoît CHÉNAIS [2], Marwa ZIDI [1], Salma DJEBBI [1], Aurore CARUSO [2], Françoise DENIS [2], Johann CONFAIS [3,4], Myriam BADAWI [2], Nathalie CASSE [2,*] and Maha MEZGHANI KHEMAKHEM [1,*]

[1]   Laboratory of Biochemistry and Biotechnology (LR01ES05), Faculty of Sciences of Tunis, University of Tunis El Manar, Tunis 1068, Tunisia; khouloud.klai@fst.utm.tn (K.K.); marwa.zidi@fst.utm.tn (M.Z.); salma.djebbi@fst.utm.tn (S.D.)

[2]   EA2160 Mer Molécules Santé, Le Mans Université, 72085 Le Mans, France; Benoit.Chenais@univ-lemans.fr (B.C.); Aurore.Caruso@univ-lemans.fr (A.C.); fdenis@univ-lemans.fr (F.D.); Myriam.Badawi@univ-lemans.fr (M.B.)

[3]   URGI, INRAE, Université Paris-Saclay, 78026 Versailles, France; johann.confais@inrae.fr

[4]   BioinfOmics, Plant Bioinformatics Facility, INRAE, Université Paris-Saclay, 78026 Versailles, France

*   Correspondence: Nathalie.Casse@univ-lemans.fr (N.C.); maha.mezghani@fst.utm.tn (M.M.K.)

Received: 29 October 2020; Accepted: 8 December 2020; Published: 11 December 2020

Simple Summary: Transposable elements (TEs) are mobile DNA sequences that can copy themselves within a host genome. TE-mediated changes in regulation can lead to massive and rapid changes in expression, responses that are potentially highly adaptive when an organism is faced with a mortality agent in the environment, such as an insecticide. *Helicoverpa armigera* shows a hight number of reported cases of insecticide resistance worldwide, having evolved resistance against pyrethroids, organophosphates, carbamates, organochlorines, and recently to macrocyclic lactone spinosad and several *Bacillus thuringiensis* toxins. In the present study, we conducted a TE annotation using combined approaches, and the results revealed a total of 8521 TEs, representing 236,132 copies, covering 12.86% of the *H. armigera* genome. In addition, we underlined TE insertions in defensome genes and we successfully identified nine TE insertions belonging to the RTE, R2, CACTA, Mariner and hAT superfamilies.

Abstract: The cotton bollworm *Helicoverpa armigera* Hübner (*Lepidoptera: Noctuidae*) is an important pest of many crops that has developed resistance to almost all groups of insecticides used for its management. Insecticide resistance was often related to Transposable Element (TE) insertions near specific genes. In the present study, we deeply retrieve and annotate TEs in the *H. armigera* genome using the Pipeline to Retrieve and Annotate Transposable Elements, PiRATE. The results have shown that the TE library consists of 8521 sequences representing 236,132 TE copies, including 3133 Full-Length Copies (FLC), covering 12.86% of the *H. armigera* genome. These TEs were classified as 46.71% Class I and 53.29% Class II elements. Among Class I elements, Short and Long Interspersed Nuclear Elements (SINEs and LINEs) are the main families, representing 21.13% and 19.49% of the total TEs, respectively. Long Terminal Repeat (LTR) and Dictyostelium transposable element (DIRS) are less represented, with 5.55% and 0.53%, respectively. Class II elements are mainly Miniature Inverted Transposable Elements (MITEs) (49.11%), then Terminal Inverted Repeats (TIRs) (4.09%). Superfamilies of Class II elements, i.e., Transib, P elements, CACTA, Mutator, PIF-harbinger, Helitron, Maverick, Crypton and Merlin, were less represented, accounting for only 1.96% of total TEs. In addition, we highlighted TE insertions in insecticide resistance genes and we successfully identified nine TE insertions belonging to RTE, R2, CACTA, Mariner and hAT superfamilies. These insertions are hosted in genes encoding cytochrome P450 (CyP450), glutathione S-transferase (GST), and ATP-binding cassette

(ABC) transporter belonging to the G and C1 family members. These insertions could therefore be involved in insecticide resistance observed in this pest.

Keywords: *Helicoverpa armigera*; transposable elements; insertions sites; insecticide resistance genes

1. Introduction

The cotton bollworm, *Helicoverpa armigera* (Hübner) (*Lepidoptera, Noctuidae*), is a serious crop pest having a worldwide distribution [1]. This polyphagous insect causes substantial damages to a wide range of hosts, including cotton, maize, sorghum, and tomato [2]. The biological and ecological traits of *H. armigera*, such as high reproduction rate, polyphagy, high mobility and facultative diapause, make it difficult to control [3]. Management of *H. armigera* attacks rely heavily on the use of chemicals [4]. However, this practice is harmful to the environment and has caused a rapid buildup of insecticide resistance in *H. armigera* populations [5].

Insecticide resistance in *H. armigera* is widespread and has evolved against most of commonly used insecticides [6]. To survive, pests have developed various mechanisms to resist against toxic compounds. These mechanisms include point mutations resulting in target-site resistance such as knockdown resistance (kdr), acetylcholinesterase (Ace-1) and receptor sub-unit termed (RDL) mutations, and also metabolic resistance with involvement of several detoxification enzymes [7].

Metabolic detoxification of toxins is the primary strategy occurring in three phases, each with its own set of enzymes or transporters. Cytochrome P450 monooxygenases (P450s) and carboxylesterases (CarE) carry out phase I, glutathione S-transferases (GSTs) and UDP-glycosyltransferases (UGTs) are phase II enzymes, and ATP-binding cassette transporters (ABC) ensure phase III [8–10]. The understanding of resistance mechanisms remains a challenge that next generation sequencing technologies and the increasing number of sequenced genomes can help to address [11].

Transposable Elements (TEs) are ubiquitous components of eukaryotic genomes that are strongly regulated and inactivated by mutations, which keep transposition events relatively rare [12,13]. However, because of their ability to replicate, TEs may accumulate in host genomes and generate abundant sites for chromosomal rearrangements, which may have deleterious or beneficial consequences [14]. In addition, TEs can provide a selective advantage through their insertion sites, which can enhance or repress gene expression or can be domesticated as new host gene [14–16]. Thus, TEs are an important source of variability for the genomes of their hosts and are therefore key to understanding their evolution. Indeed, TEs may be involved in the genetic adaptation of organisms such as insects to stressful environments, among which is the acquisition of insecticide resistance [17].

Several studies have shown that insecticide resistance can be associated with TE insertions in specific genes. For example, dichlorodiphenyltrichloroethane (DTT) resistance in *Drosophila melanogaster* was correlated with the insertion of a Long Terminal Repeat (LTR)-gypsy retroelement into the 5′ region of the cytochrome P450 gene [18,19]. In *Helicoverpa zea*, several TE insertions in regulatory regions, exons and introns of cytochrome P450 genes were related with pyrethroid resistance and xenobiotic metabolism [20].

TEs are classified into two major classes depending on the transposition intermediate. Class I, or retroelements, replicate and transpose via an RNA intermediate; while Class II elements, or DNA transposons, are mobilized via a DNA intermediate [21]. According to the classification of Wicker et al. [22], each class is subdivided into orders and superfamilies. Class I elements are further subdivided into long terminal repeat (LTR) retrotransposons and non-LTR retrotransposons. The non-LTR retrotransposons include the long interspersed nuclear elements (LINEs) and the short interspersed nuclear elements (SINEs) as well as the Penelope-Like Elements (PLEs). Class II elements are subdivided into two subclasses. Subclass 1 includes the terminal inverted repeat (TIR) transposons and Crypton-like elements, which cleave both DNA intermediate strands, while subclass 2 elements including Mavericks and Helitrons with a single-strand DNA intermediate have a replicative mode of transposition [22–24]. Class II TIRs transposons also include Miniature Inverted-repeat Transposable Elements (MITEs), which are short (~100 to 800 bp) non-autonomous truncated versions of autonomous transposable elements. MITEs possess conserved terminal inverted repeats (TIRs ≥ 10 bp) and a target site duplication (TSDs = 2~10 bp) [25,26].

Annotation of TEs is a challenging task because of their diversity, their repetitive nature and the complexity of their structures, and numerous tools have been designed to identify TEs [27]. In this study, we used the Pipeline to Retrieve and Annotate Transposable Elements (PiRATE) [28] to annotate the mobilome of *H. armigera* and pinpoint TEs inserted in defensome genes.

2. Material and Methods

2.1. Mobilome Annotation

The *H. armigera* genome available in GenBank-NCBI (BioProject PRJNA378437) is 337,087 Mb. This genome is assembled in 24,552 contigs and 998 scaffolds corresponding to 23.5 kb and 1000 kb N50 length respectively [29].

H. armigera genome assembly and the corresponding raw Illumina data were both submitted to the PiRATE pipeline to search for TEs following three steps [28]. In the first step, putative TE sequences were detected using four approaches. The first approach is a similarity-based detection of TEs using RepeatMasker [30] and TE-HMMER [31]. The second approach is a structure-based detection, using LTRharvest [32], MGEScan-nonLTR [33] and SINE-Finder [34]. The repetitiveness-based detection is the third approach using TE *de novo* [35] and Repeat Scout [36]. The last approach is a *de novo* approach using the dnaPipeTE tool [37].

After TE detection, a second step was performed to eliminate redundant sequences and classify the remaining sequences using the PASTEC tool following the Wicker's 80-80-80 rules corresponding to sequences longer than 80 bp, sharing more than 80% sequence identity and over 80% of their length [22,38]. Two libraries were generated: a "total TEs library" containing the potentially autonomous TEs and the non-autonomous TEs, and a "repeated elements library" containing the uncategorized repeated sequences and the non-TE sequences. Subsequently, two runs of TEannot [35] were performed for each library to generate the final libraries of total TEs and total repeats. To refine the annotation of TE copies in the whole genome, we used the TEannot pipeline from the REPET package v3.0 with TEs sequences of PiRATE step 1 that align at least with one Full-Length Copy (FLC) on the genome assembly [35,39,40].

Finally, a manual curation was released for all annotated TEs to find corresponding families. This analysis was performed by nucleotide Basic Local Alignment Search Tool (BLAST) against Repbase (48,225 TE sequences) and Dfam (6959 sequences) databases using a threshold value of 80% (Figure 1).

To identify putative MITE sequences, the *H. armigera* genome assembly was submitted to the MITE Tracker tool [41] (Figure 1). This tool searches for putative inverted repeat sequences ranging from 50 to 800 bp. Subsequently, putative MITEs were aligned and clustered into families by Vsearch [42] based on target sites duplication (TSD) and Terminal Inverted Repeat (TIR) sequences.

2.2. Search for TE Insertions in Defensome Genes

Annotated TEs from the *H. armigera* genome have been extended by 50 kb both upstream and downstream of their DNA sequences. The nucleotide BLAST was used to find defensome genes in the extended regions using 80% similarity and 80% query coverage threshold (Figure 1).

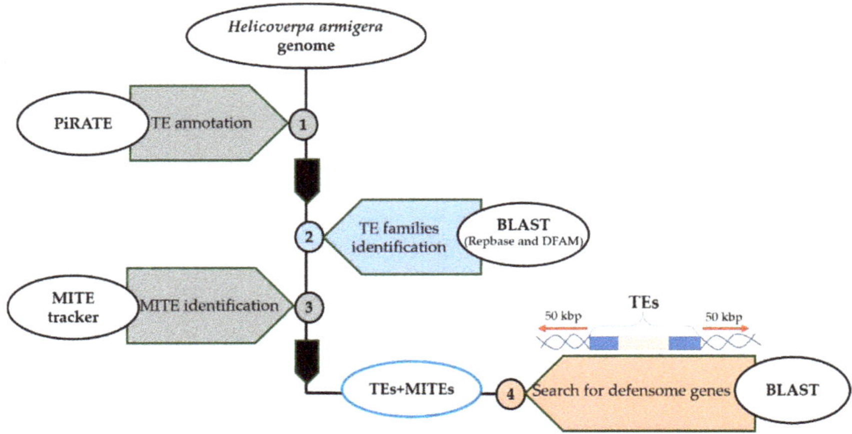

Figure 1. Flowchart for transposable elements annotation and insertion sites identification in the *H. armigera* genome.

3. Results

3.1. TEs Annotation of the H. armigera Genome

The screening of TEs in the *H. armigera* genome using different detection tools led to the identification of 100,184 TE candidates after redundance elimination. The classification step generated two libraries: the "total TE library" and the "total repeats library" containing 4336 sequences (11.36%) and 5201 sequences (13.63%), respectively. Among the TE library, 3133 sequences were identified as FLC belonging to Class I (2720 FLC), Class II (413 FLC) and 61 sequences were undefined (Supplementary File 1). A total of 70,030 sequences was classified as non-TE (41,411 sequences) and unclassified TEs (28,619 sequences). The results revealed that TE sequences cover 12.86% (43,349,853 bp) of the *H. armigera* genome and most of the TE sequences belong to Class II elements, accounting for 53.29% of the total TE content, while Class I elements account for 46.71%.

Among Class I elements, SINEs and LINEs were the main families, representing 21.13% and 19.49% of the total TEs, respectively. LTR elements were represented with 5.55% and Dictyostelium transposable element (DIRS) with only 0.53%. The Class II elements were represented mainly by MITEs and TIRs with 49.11% and 4.09%, respectively.

To investigate the evolutionary history of TEs in the *H. armigera* genome, we plotted the distribution of identity values between copies and their representative sequences. Distributions of TE classes showed a peak at 80% identity for Class I elements, while, for Class II elements, the distribution was linear with a recent burst at 98% identity (Figure 2).

This analysis revealed that the *H. armigera* genome has undergone a multitude of ancient and recent bursts of different TE superfamilies showing its fluidity. Distributions of TE copies showed three peaks of transposition activity (Figure 3). The first peak is at 65% divergence involving a burst of the DIRS order. We also noted a second TE burst, particularly for LINE, SINE, LTR, TIR, and Helitron orders at 80% identity. In addition, the distribution also showed the appearance of MITE elements at 95% identity, suggesting a recent invasion of the *H. armigera* genome by these TEs (Figure 3).

3.1.1. Class I Retrotransposons

The annotation of Class I retrotransposons in the *H. armigera* genome allowed for the identification of 3980 sequences representing 186,645 copies belonging to 10 superfamilies (Table 1).

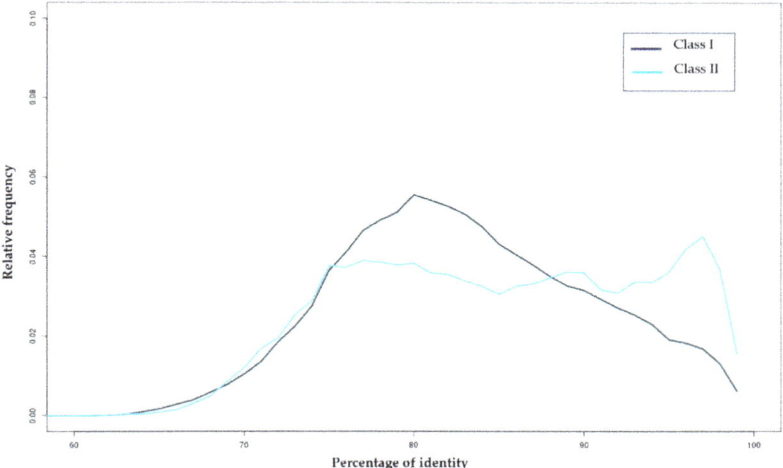

Figure 2. Distribution of sequence identity values between TE copies and TE sequences with at least one full-length copy for Class I and Class II elements. The relative frequencies per percentage of identity of Class I and Class II are represented in different colors.

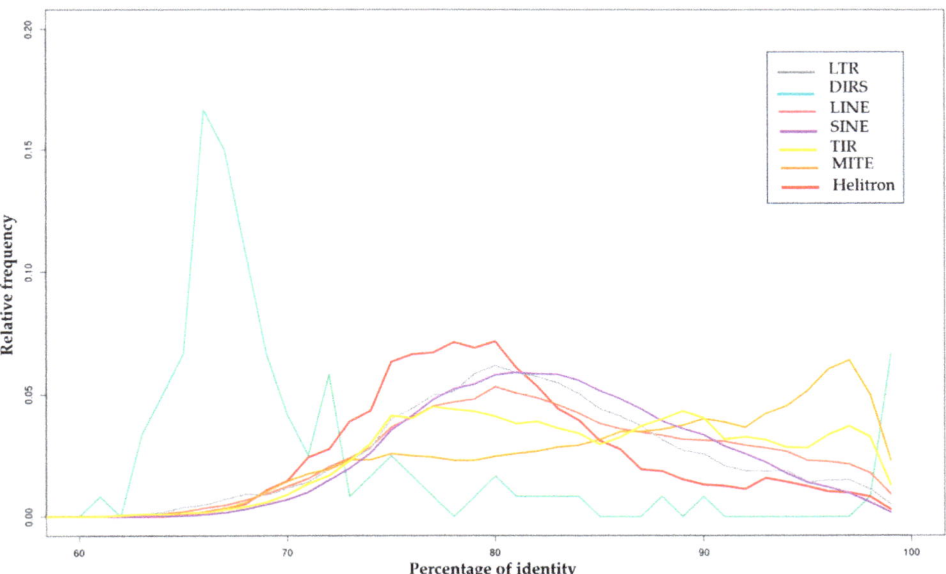

Figure 3. Distribution of sequence identity values between TE copies and TE sequences with at least one FLC. The relative frequencies per percentage of identity of Dictyostelium transposable element (DIRS), Helitron, Long and Short Interspersed Nuclear Elements (LINE and SINE), Long Terminal Repeat (LTR), Miniature Inverted Transposable Element (MITE) and Terminal Inverted Repeat (TIR) orders are represented in different colors. Only the main orders (in terms of copy number) are represented.

From the LTR retrotransposons, 345 TEs were full-length copies (FLC). Gypsy was the most abundant LTR superfamily with 241 sequences and 14,876 copies followed by Bel-Pao (155 sequences) and Copia (77 sequences) with 97 and 67 FLC, respectively. BLAST searches against Repbase and Dfam databases showed that, for all identified LTR sequences, no similarity was found with TEs in databases.

Table 1. Summary of the identified and annotated TEs in the *H. armigera* genome.

Class	Order	Superfamily	Total Sequences [1]	Full-Length Copies [2]	TE Percentage	Copy Number [3]
	LTR	Gypsy	241	181		14,876
		Bel pao	155	97		1792
		Copia	77	67		223
	Total LTR		473	345	5.55%	16,891
	DIRS	DIRS	45	38		149
Class I	**Total DIRS**		45	38	0.53%	149
	LINE	Jockey	716	480		44,432
		RTE	604	543		46,458
		I	172	93		16,520
		R2	169	101		10,614
	Total LINE		1661	1217	19.49%	118,024
	SINE	tRNA	1797	1120		51,581
		5S	4	-		0
	Total SINE		1801	1120	21.13%	51,581
	Total Class I		3980	2720	46.71%	186,645
	TIR	hAT	120	105		6267
		Mariner	106	74		2284
		Piggybac	41	32		6670
Class II Subclass I		Transib	40	32		188
		P	15	12		6103
		CACTA	14	11		201
		Mutator	6	4		477
		PIF_harbinger	6	5		90
		Merlin	1	1		30
	Total TIR		349	276	4.09%	22,310
Class II Subclass II	Crypton	Crypton	1	1		3
	Helitron	Helitron	4	4		595
	Maverick	Maverick	2	2		14
	MITEs	MITEs	4185	130	49.11%	26,565
	Total classII		4541	413	53.29%	49,487
	Total TEs		8521	3133	100	236,132

[1] Representative sequence identified with PiRATE Step1 with an identity ≤ 80% [2] TEs sequences of PiRATE step 1 that align at least with one Full-length Copy (FLC) on the genome assembly [3] Copies annotated with TEannot REPET package v3.0. Number of undefined transposable elements are not shown in the table.

Regarding LINEs retrotransposons, the most abundant elements belong to the Jockey and RTE superfamilies, with 716 and 604 sequences, respectively, representing 44,432 and 46,458 copies, respectively. The RTE superfamily contains the highest number of FLCs among all annotated TEs (Table 1). According to the BLAST searches, no similarity was found for Jockey elements in databases while 17 RTE sequences showed similarity ranging from 86% to 100% with RTE-1_Avan, RTE-2_Hmel_C, RTE-3_DPl, RTE-4_DPl, RTE-5_DPl and Proto2-1_BM families (Table S1).

Concerning SINEs retrotransposons, the tRNA-derived SINE superfamily corresponds to 45.15% of all Class I TEs with 1797 sequences representing 51,581 copies and 1120 FLCs. Among these sequences, 578 SINEs belong to the HaSE1 family while 85 sequences fell into the HaSE3 family with a similarity ranging from 80% to 100% (Table S1).

3.1.2. Class II Transposons

Our results revealed that DNA transposons in the *H. armigera* genome are represented by a total of 4541 sequences representing 49,487 copies belonging to 13 superfamilies (Table 1).

Regarding TIR elements, the hAT superfamily was the most abundant, including 120 sequences, among which 105 elements were FLC. BLAST searches against Repbase and Dfam databases revealed high similarity of three hAT elements with the hAT-1_DAN family (Table S1).

The Tc1/mariner superfamily was represented by 106 elements (2284 copies), among which 74 FLC. Research using BLAST has shown that 12 of these elements are distributed among five families—ANM4, nMar-2_Avan, nMar-18_Hmel, Mariner-1_AMel and Mariner-3_BM—exhibiting similarity ranging from 80.25% to 95.34% (Table S1).

A total of 41 sequences (6670 copies) belonging to the PiggyBac superfamily was identified in the *H. armigera* genome. Further analysis revealed that only four among the identified sequences showed similarity ranging from 81% to 99% with already identified PiggyBac transposons, npiggyBac-8 and piggyBac-2.

The following superfamilies of Class II elements, i.e., Transib, P elements, CACTA, Mutator, PIF-harbinger, Helitron, Maverick, Crypt-on and Merlin, were less represented in the *H. armigera* genome corresponding to a total of 7701 copies.

In addition, the MITE tracker allowed for the identification of 4185 putative MITEs corresponding to 43.11% of all annotated TEs. The analysis of terminal TIRs and TSD sequences allowed for the classification of 3570 MITE sequences (26,565 copies) into seven superfamilies (Table 2). The Tc1\mariner and PIF-Harbinger superfamilies were the most represented, with 1817 and 1368 MITEs, respectively, followed by the CACTA superfamily with 250 MITEs, then PiggyBac with 93 elements. The hAT, Transib and Maverick superfamilies were represented by only 20, 16 and six MITEs, respectively.

Table 2. Distribution of MITEs identified in the *H. armigera* genome.

Superfamily	TSD	Number of MITEs	MITE Length (bp)	TIR Length (bp)
Tc1/mariner	TA	1817	50–360	10–21
PIF-Harbinger	TWA	1368	55–685	15–32
CACTA	2–3 bp	250	78–775	10–26
Piggybac	TTAA	93	50–800	15–31
hAT	T****A	20	56–260	17–29
Transib	C***G	16	83–386	13–27
Maverick	6 bp	6	50–800	10–24
Other	-	615	50–800	11–35

3.2. TE Insertions Scanning in Defensome Genes

Nucleotide BLAST searches for defensome genes in the regions framing identified TE sequences led to the identification of nine TE insertion sites in seven genes encoding for detoxifying enzymes (Table 3). The involved TEs are members of RTE, R2, CACTA, DIRS, Mariner, and hAT superfamilies (Table 3). Further analysis of TE insertions has shown that five TEs were inserted in four cytochrome P450 genes, an element was retrieved in a GST gene, two were hosted by ABC-G transporter gene and one was inserted in an ABC-C1 transporter gene.

Table 3. TE insertions in genes encoding for detoxifying enzymes in *H. armigera*.

Gene Family	Gene Name	Gene Length (bp)	Inserted Element Name	TE Length (bp)	Insertion Position
Cytochrome P450 (CYP450)	4g15-like (LOC110375617)	14,974	LINE_R2_526	4083	6483–10,566
	4C1-like (LOC113006340)	39,707	LINE_RTE17512	620	5655–6274
			LINE_RTE84107	740	7021–7763
	4C1-like (LOC110381376)	8332	LTR_CACTA3565	1179	6899–8066
	6k1-like (LOC110381042)	15,958	LINE_jockey2199	1629	9431–10,438
Glutathione S-transferase (GST)	GST 1-like (LOC110371343)	4001	TIR_Mariner2770	1304	1793–3097
ATP binding cassette (ABC) transporter	ABC-G member 20 (LOC110376033)	96,146	TIR_Mariner419	1085	91,998–93,071
			TIR_hAT2824	951	1486–2436
	ABC-C1 (Multidrug resistance protein homolog 49-like MRP1) (LOC110377844)	48,302	LINE_RTE63004	894	38,590–39,483

Six of the inserted TEs have intronic insertion sites and one TE insertion occurred in the first exon of the ABCG transporter member 20 gene (Figure 4 and Figures S1–S7). It should be noted that, for the cytochrome P450 4C1-like (LOC11300634) gene, harboring two LINE insertions, no exon or intron information was retrieved in GenBank.

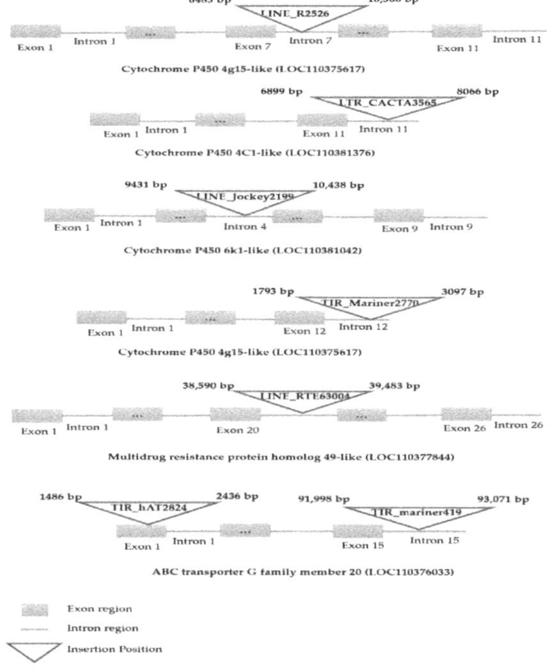

Figure 4. Schematic representation of TE insertion sites in genes encoding for detoxifying enzymes in *H. armigera*.

4. Discussion

The present study identified TEs present in the genome of *H. armigera* and searched for their occurrence among the defensome genes of this pest. Characterizing TEs is an important task for non-model organisms and several TE annotation tools have recently been developed for TE characterization in these organisms [43]. In this work, we used the PiRATE pipeline [28], which combines different TE identification tools to detect, classify and annotate TEs into known superfamilies.

The results revealed a total TE content of 4336 sequences, covering 12.86% of the *H. armigera* assembly which is much higher than the previous data from Pearce et al. [29], in which Repeat Masker and Repeat Modeller tools allowed for the identification of a TE content representing only 0.88% of the genome. This confirms the interest of using a pipeline like PiRATE to increase the detection of TEs in a genome.

In other lepidopteran genomes, such as *Bombyx mori* and *Spodoptera frugiperda*, about 53% of TEs were previously identified as Class I and half of these belong to the *LINE* order [44]. Consistent with these results, 46.71% of the TEs identified in the *H. armigera* genome were Class I elements and the two most abundant were the *SINE* and *LINE* orders. These results suggest that using a combined approach is more specific for short TEs than the use of a single tool.

Some TEs were previously identified in *H. armigera* by molecular biology methods. In 2008, Sun et al. identified two *PiggyBac* elements (HaPLE1 and HaPLE2) in the cotton bollworm genome [45]. A few years later, multiple copies of two distinct mariner elements, *Hamar1* and *Hamar2*, were isolated by Wang et al. [46]. All these sequences were retrieved in the current study with a minimum of 88% identity.

To pinpoint TE insertions in defensome genes, the upstream and downstream regions of TE sequences were scanned and nine insertions have been successfully identified. Among them, five were retrieved in cytochrome P450 genes introns. Cytochrome P450 are among the three main groups of detoxification enzymes used by insects that play crucial roles in the detoxification of most pesticides [47]. In *H. armigera*, two *LINEs* (RTE and R2) and a *CACTA* element were inserted in the P450 4C1-like gene, the expression of which is responsible for the insecticide detoxification [48]. In *Aphis gossypii*, another LINE, the *Jockey* element was also identified in cytP450 6K1-like gene intron, putting it as the strongest candidate conferring resistance to thiamethoxam [49]. Two other *LINE* elements were inserted into the P450 4g15-like gene, but no information for exon or intron regions was retrieved in GenBank for this gene, suggesting that the *H. armigera* genome needs more annotation. However, the expression of this gene in *Aphis gossypii* was correlated with imidacloprid resistance [50].

Arthropod ABC transporters also play an important role in metabolite detoxification [10]. Three TEs, which belong to *RTE*, *mariner* and *hAT* superfamilies, were inserted in an ABCG transporter and the gene in the *H. armigera* genome. TE insertions into non-coding regions may be less subject to selection and could be influenced by other mechanisms of control. These insertions could be successfully spliced out during mRNA processing and thus may have no obvious effects on the function of the corresponding defensome gene [51]. Alternatively, they could result in exon skipping, alternative splicing, or alterations in expression profiles if the corresponding IR gene introns contain regulatory sequences, as exemplified by the insertion of the *Mu* element into an intron of the *Knotted* locus in Maize [51]. Analyses of the transcript and the expression levels of the identified IR genes as well as toxicity bioassay will be necessary to determine the exact effects of the nine identified TE insertions on the expression and function of the identified IR genes, as well as on the fitness of this species in the presence of a stressful environment such as insecticides.

5. Conclusions

Genome-wide TE annotation has rarely been performed in *H. armigera*. This study opens the way to new searches about the role of TEs in the genome evolution of *H. armigera* and their contribution to the pest adaptation such as insecticide resistance. In the present study, we conducted an accurate TE annotation, and the results reveal a total of 8521 TEs covering 12.86% (43349853 bp) of the *H. armigera* genome. The annotation of TEs was crossed with insertion sites search in defensome genes. Nine TEs belonging to the RTE, R2, LTR and TIR superfamilies were found to be inserted in CYP450, GST and ABC transporter genes and their insertion sites were mostly in intronic regions except for a hAT element inserted in the exon region. These results present for the first-time evidence that TEs are present in IR genes of *H. armigera*. However, further studies are necessary to elucidate the functional relationship of TEs with IR genes in *H. armigera*.

Supplementary Materials: The following are available online at http://www.mdpi.com/2075-4450/11/12/879/s1. File 1. Full-length TEs annotated in the *Helicoverpa armigera* genome, Table S1: BLAST results of annotated TEs in *H. armigera* against reference TEs in databases (Repbase and Dfam), Figure S1. Alignment of LINE_R2526 element with Cytochrome P450 4g15-like gene, Figure S2. Alignment of LINE_RTE84107 and LINE_RTE17512 elements with Cytochrome P450 4C-like gene, Figure S3. Alignment of TIR_CACTA3565 element with Cytochrome P450 4C-like gene, Figure S4. Alignment of LINE_Jockey2199 element with Cytochrome P450 6k1-like gene, Figure S5. Alignment of TIR_mariner2770 element with glutathione S-transferase 1-like gene, Figure S6. Alignment of LINE_RTE63004 element with ABC-C1 homolog 49-like gene, Figure S7. Alignment of TIR_mariner419 and TIR_hAT824 elements with ABCG member 20 gene.

Author Contributions: Conceptualization, K.K., M.M.K., A.C., B.C. and N.C.; methodology, K.K., M.Z., M.B., J.C. and F.D.; investigation, data curation, K.K.; writing—original draft preparation, K.K., M.M.K., B.C. and N.C.; writing—review and editing, M.M.K., B.C., N.C., A.C., S.D., F.D.; supervision, M.M.K., B.C. and N.C.; project administration, M.M.K. and N.C.; funding acquisition, M.M.K. and N.C. All authors have read and agreed to the published version of the manuscript.

Funding: This research was funded by PHC-Utique "Hubert Curien Franco-Tunisian partnership" (41703ZJ) and the Tunisian Ministry of Higher Education and Scientific Research and the University of Tunis El Manar.

Conflicts of Interest: The authors declare no conflict of interest.

References

1. Burgio, G.; Ravaglia, F.; Maini, S.; Bazzocchi, G.G.; Masetti, A.; Lanzoni, A. Mating Disruption of *Helicoverpa armigera* (Lepidoptera: Noctuidae) on Processing Tomato: First Applications in Northern Italy. *Insects* **2020**, *11*, 206. [CrossRef] [PubMed]

2. Tembrock, L.R.; Timm, A.E.; Zink, F.A.; Gilligan, T.M. Phylogeography of the Recent Expansion of *Helicoverpa armigera* (Lepidoptera: Noctuidae) in South America and the Caribbean Basin. *Ann. Entomol. Soc. Am.* **2019**, *112*, 388–401. [CrossRef]

3. Sarate, P.J.; Tamhane, V.A.; Kotkar, H.M.; Ratnakaran, N.; Susan, N.; Gupta, V.S.; Giri, A.P. Developmental and digestive flexibilities in the midgut of a polyphagous pest, the cotton bollworm, *Helicoverpa armigera*. *J. Insect Sci.* **2012**, *12*, 42. [CrossRef] [PubMed]

4. Pereira, F.P.; Reigada, C.; Diniz, A.J.F.; Parra, J.R.P. Potential of Two Trichogrammatidae species for *Helicoverpa armigera* control. *Neotrop. Entomol.* **2019**, *48*, 966–973. [CrossRef]

5. Ansari, M.S.; Moraiet, M.A.; Ahmad, S. Insecticides: Impact on the Environment and Human Health. In *Environmental Deterioration and Human Health: Natural and Anthropogenic Determinants*; Malik, A., Grohmann, E., Akhtar, R., Eds.; Springer: Dordrecht, The Netherlands, 2014; pp. 99–123. [CrossRef]

6. Srinivas, R.; Udikeri, S.S.; Jayalakshmi, S.K.; Sreeramulu, K. Identification of factors responsible for insecticide resistance in *Helicoverpa armigera*. *Comp. Biochem. Physiol.* **2004**, *137*, 261–269. [CrossRef]

7. Sarwar, M.; Salman, M. Insecticides Resistance in Insect Pests or Vectors and Development of Novel Strategies to Combat Its Evolution. *IJBBE* **2015**, *1*, 344–351.

8. Sampath, P.; Murukarthick, J.; Izzah, N.K.; Lee, J.; Choi, H.I.; Shirasawa, K.; Choi, B.S.; Liu, S.; Nou, I.S.; Yang, T.J. Genome-Wide Comparative Analysis of 20 Miniature Inverted-Repeat Transposable Element Families in *Brassica rapa* and *B. oleracea*. *PLoS ONE* **2014**, *9*, e94499. [CrossRef]

9. De Marco, L.; Sassera, D.; Epis, S.; Mastrantonio, V.; Ferrari, M.; Ricci, I.; Comandatore, F.; Bandi, C.; Porretta, D.; Urbanelli, S. The choreography of the chemical defensome response to insecticide stress: Insights into the *Anopheles stephensi* transcriptome using RNA-Seq. *Sci. Rep.* **2017**, *7*. [CrossRef]

10. Dermauw, W.; van Leeuwen, T. The ABC gene family in arthropods: Comparative genomics and role in insecticide transport and resistance. *Insect Biochem. Mol. Biol.* **2014**, *45*, 89–110. [CrossRef]

11. Su, H.; Gao, Y.; Liu, Y.; Li, X.; Liang, Y.; Dai, X.; Xu, Y.; Zhou, Y.; Wang, H. Comparative transcriptome profiling reveals candidate genes related to insecticide resistance of *Glyphodes pyloalis*. *Bull. Entomol. Res.* **2020**, *110*, 57–67. [CrossRef]

12. Mérel, V.; Boulesteix, M.; Fablet, M.; Vieira, C. Transposable elements in *Drosophila*. *Mob. DNA* **2020**, *11*, 23. [CrossRef] [PubMed]

13. Lavoie, C.A. Transposable Element Content in Non-Model Insect Genomes. Master's Thesis, Mississippi State University, Starkwell, MS, USA, 2020.

14. Chénais, B.; Caruso, A.; Hiard, S.; Casse, N. The impact of transposable elements on eukaryotic genomes: From genome size increase to genetic adaptation to stressful environments. *Gene* **2012**, *509*, 7–15. [CrossRef] [PubMed]

15. Bourgeois, Y.; Boissinot, S. On the Population Dynamics of Junk: A Review on the Population Genomics of Transposable Elements. *Genes* **2019**, *10*, 419. [CrossRef] [PubMed]

16. Steinberg, C.E.W. *Arms Race between Plants and Animals: Biotransformation System in Stress Ecology: Environmental Stress as Ecological Driving Force and Key Player in Evolution*; Springer: New York, NY, USA, 2012; pp. 61–105.

17. Le Goff, G.; Hilliou, F. Resistance evolution in *Drosophila*: The case of CYP6G1. *Pest Manag. Sci.* **2017**, *73*, 493–499. [CrossRef]

18. Catania, F.; Kauer, M.O.; Daborn, P.J.; Yen, J.L.; French-Constant, R.H.; Schlötterer, C. World-wide survey of an Accord insertion and its association with DDT resistance in *Drosophila melanogaster*. *Mol. Ecol.* **2004**, *13*, 2491–2504. [CrossRef]

19. Schlenke, T.A.; Begun, D.J. Strong selective sweep associated with a transposon insertion in *Drosophila simulans*. *Proc. Natl. Acad. Sci. USA* **2004**, *101*, 1626–1631. [CrossRef]

20. Chen, S.; Li, X. Transposable elements are enriched within or in close proximity to xenobiotic-metabolizing cytochrome P450 genes. *BCM Evol. Biol.* **2007**, *7*, 46. [CrossRef]

21. Finnegan, D.J. Eukaryotic transposable elements and genome evolution. *Trends Genet.* **1989**, *5*, 103–107. [CrossRef]

22. Wicker, T.; Sabot, F.; Hua-Van, A.; Bennetzen, J.L.; Capy, P.; Chalhoub, B.; Flavell, A.; Leroy, P.; Morgante, M.; Panaud, O.; et al. An unified classification system for eukaryotic transposable elements. *Nat. Rev. Genet.* **2007**, *8*, 973–982. [CrossRef]

23. Kapitonov, V.V.; Jurka, J. Rolling-Circle Transposons in Eukaryotes. *Proc. Natl. Acad. Sci. USA* **2001**, *98*, 8714–8719. [CrossRef]

24. Pritham, E.J.; Putliwala, T.; Feschotte, C. Mavericks, a novel class of giant transposable elements widespread in eukaryotes and related to DNA viruses. *Gene* **2007**, *390*, 3–17. [CrossRef] [PubMed]

25. Kuang, H.; Padmanabhan, C.; Li, F.; Kamei, A.; Bhaskar, P.B.; Ouyang, S.; Jiang, J.; Robin Buell, C.; Baker, B. Identification of miniature inverted-repeat transposable elements (MITEs) and biogenesis of their SiRNAs in the Solanaceae: New functional implications for MITEs. *Genome Res.* **2009**, *19*, 42–56. [CrossRef] [PubMed]

26. Schrader, L.; Schmitz, J. The impact of transposable elements in adaptive evolution. *Mol. Ecol.* **2019**, *28*, 1537–1549. [CrossRef] [PubMed]

27. Yan, H.; Torchiana, F.; Bombarely, A. Guideline for genome transposon annotation derived from evaluation of popular TE identification tools. *Preprints* **2020**, 2020080275. [CrossRef]

28. Berthelier, J.; Casse, N.; Daccord, N.; Jamilloux, V.; Saint-Jean, B.; Carrier, G. A transposable element annotation pipeline and expression analysis reveal potentially active elements in the microalga *Tisochrysis lutea*. *BMC Genom.* **2018**, *19*, 1–14. [CrossRef]

29. Pearce, S.L.; Clarke, D.F.; East, P.D.; Elfekih, S.; Gordon, K.H.J.; Jermiin, L.S.; McGaughran, A.; Oakeshott, J.G.; Papanikolaou, A.; Perera, O.P.; et al. Genomic innovations, transcriptional plasticity and gene loss underlying the evolution and divergence of two highly polyphagous and invasive *Helicoverpa* pest species. *BMC Biol.* **2017**, *15*, 1–30. [CrossRef]

30. Smit, A.F.A.; Hubley, R.; Green, P. Repeat Masker. Unpublished Data. Current Version: Open-4.09 (Dfam:3.0 only) 1996. Available online: http://www.repeatmasker.org (accessed on 15 April 2019).

31. Eddy, S.R. Multiple alignment using hidden Markov models. *ISMB* **1996**, *3*, 114–120.

32. Ellinghaus, D.; Kurtz, S.; Willhoeft, U. LTR harvest, an efficient and flexible software for de novo detection of LTR retrotransposons. *BMC Bioinform.* **2008**, *9*, 18. [CrossRef]

33. Rho, M.; Tang, H. MGEScan-Non-LTR: Computational identification and classification of autonomous non-LTR retrotransposons in eukaryotic genomes. *Nucleic Acids Res.* **2009**, *37*, e143. [CrossRef]

34. Wenke, T.; Döbel, T.; Sörensen, T.R.; Junghans, H.; Weisshaar, B.; Schmidta, T. Targeted Identification of Short Interspersed Nuclear Element Families Shows Their Widespread Existence and Extreme Heterogeneity in Plant Genomes. *Plant Cell* **2011**, *23*, 3117–3128. [CrossRef]

35. Flutre, T.; Duprat, E.; Feuillet, C.; Quesneville, H. Considering Transposable Element Diversification in de Novo Annotation Approaches. *PLoS ONE* **2011**, *6*, e16526. [CrossRef] [PubMed]

36. Price, A.L.; Jones, N.C.; Pevzner, P.A. De novo identification of repeat families in large genomes. *Bioinformatics* **2005**, *21* (Suppl. 1), i351–i358. [CrossRef] [PubMed]

37. Goubert, C.; Modolo, L.; Vieira, C.; Moro, C.V.; Mavingui, P.; Boulesteix, M. De Novo Assembly and Annotation of the Asian Tiger Mosquito (*Aedes albopictus*) Repeatome with DnaPipeTE from Raw Genomic Reads and Comparative Analysis with the Yellow Fever Mosquito (*Aedes aegypti*). *GBE* **2015**, *7*, 1192–1205. [CrossRef] [PubMed]

38. Hoede, C.; Arnoux, S.; Moisset, M.; Chaumier, T.; Inizan, O.; Jamilloux, V.; Quesneville, H. PASTEC: An Automatic Transposable Element Classification Tool. *PLoS ONE* **2014**, *9*, e91929. [CrossRef]

39. Quesneville, H.; Bergman, C.M.; Andrieu, O.; Autard, D.; Nouaud, D.; Ashburner, M.; Anxolabehere, D. Combined Evidence Annotation of Transposable Elements in Genome Sequences. *PLoS Comput. Biol.* **2005**, *1*, 0166–0175. [CrossRef]

40. Daccord, N.; Celton, J.M.; Linsmith, G.; Becker, C.; Choisne, N.; Schijlen, E.; van de Geest, H.; Bianco, L.; Micheletti, D.; Velasco, R.; et al. High-quality de novo assembly of the apple genome and methylome dynamics of early fruit development. *Nat. Genet.* **2017**, *49*, 1099–1106. [CrossRef]

41. Crescente, J.M.; Zavallo, D.; Helguera, M.; Vanzetti, L.S. MITE Tracker: An accurate approach to identify miniature inverted-repeat transposable elements in large genomes. *BMC Bioinform.* **2018**, *19*, 348. [CrossRef]

42. Rognes, T.; Flouri, T.; Nichols, B.; Quince, C.; Mahé, F. VSEARCH: A versatile open source tool for metagenomics. *PeerJ* **2016**, *4*, 2584. [CrossRef]

43. Goerner-Potvin, P.; Bourque, G. Computational tools to unmask transposable elements. *Nat. Rev. Genet.* **2018**, *19*, 688–704. [CrossRef]

44. D'Alençon, E.; Sezutsu, H.; Legeai, F.; Permal, E.; Bernard-Samain, S.; Gimenez, S.; Gagneur, C.; Cousserans, F.; Shimomura, M.; Brun-Barale, A.; et al. Extensive synteny conservation of holocentric chromosomes in lepidoptera despite high rates of local genome rearrangements. *Proc. Natl. Acad. Sci. USA* **2010**, *107*, 7680–7685. [CrossRef]

45. Sun, Z.C.; Wu, M.; Miller, T.A.; Han, Z.J. PiggyBac-like elements in cotton bollworm, *Helicoverpa armigera* (Hübner). *Insect Mol. Biol.* **2008**, *17*, 9–18. [CrossRef] [PubMed]

46. Wang, J.; Miller, T.A.; Park, Y. Identification of mariner-like elements belonging to the Cecropia subfamily in two closely related *Helicoverpa* species. *Insect Sci.* **2011**, *18*, 619–628. [CrossRef]

47. Zhu, W.; Yu, R.; Wu, H.; Zhang, X.; Liu, Y.; Zhu, K.Y.; Zhang, J.; Ma, E. Identification and characterization of two CYP9A genes associated with pyrethroid detoxification in *Locusta migratoria*. *Pestic. Biochem. Physiol.* **2016**, *132*, 65–71. [CrossRef] [PubMed]

48. Gong, J.; Cheng, T.; Wu, Y.; Yang, X.; Feng, Q.; Mita, K. Genome-wide patterns of copy number variations in *Spodoptera litura*. *Genomics* **2019**, *111*, 1231–1238. [CrossRef]

49. Langfield, K.L. Characterization of Neonicotinoid Resistance in the Cotton Aphid, *Aphis gossypii* from Australian Cotton. Ph.D. Thesis, University of Technology Sydney, Sydney, Australia, April 2017.

50. Kim, J.I.; Kwon, M.; Kim, G.H.; Kim, S.Y.; Lee, S.H. Two mutations in nAChR beta subunit is associated with imidacloprid resistance in the *Aphis gossypii*. *J. Asia Pac. Entomol.* **2015**, *18*, 291–296. [CrossRef]

51. Greene, B.; Wako', R.; Hake, S. Mutator Insertions in an Intron of the Maize Knotted1 Gene Result in Dominant Suppressible Mutations. *Genetics* **1994**, *138*, 1275–1285.

Publisher's Note: MDPI stays neutral with regard to jurisdictional claims in published maps and institutional affiliations.

Article

A Dynamic Energy Budget Approach for the Prediction of Development Times and Variability in *Spodoptera frugiperda* Rearing

Andre Gergs * and **Christian U. Baden**

Bayer AG, Crop Science Division, Alfred-Nobel Straße 50, 40789 Monheim, Germany; christian.baden@bayer.com
* Correspondence: andre.gergs@bayer.com

Simple Summary: The fall armyworm *Spodoptera frugiperda* is a moth that is active during the night. Its larvae cause extensive damage to many crops. Laboratory experiments are conducted to find effective ways to control this pest. One important aspect in research, generally, is that experimental results are reproducible. Reproducibility directly depends on the homogeneity of the test material—the fall armyworm larvae, in our case. The more variable the conditions of the larvae in terms of larval stages or sizes, the more variable and less reliable the research results will be. We used a mathematical model to explore the causes for increased variability in the larval development of the fall armyworm. We found that low air temperatures and poor nutrition increase development times and variability compared to higher air temperature settings and good-quality food. This finding helps researchers to adjust rearing temperatures in a way that allows starting experiments with specific larval stages and low variability on time as planned for their high-quality research.

Abstract: A major challenge in insect rearing is the need to provide certain life cycle stages at a given time for the initiation of experimental trials. The timing of delivery, organism quality, and variability directly affect the outcome of such trials. Development times and intraspecific variability are directly linked to the availability of food and to the ambient temperature. Varying temperature regimes is an approach to adapt development times to fulfill experimental needs without impairment of larval quality. However, current practices of temperature setting may lead to increased variability in terms of development times and the frequency of particular life stages at a given point in time. In this study, we analyzed how resource availability and ambient temperature may affect the larval development of the economically important noctuid species *Spodoptera frugiperda* by means of dynamic energy budget modeling. More specifically, we analyzed how rearing practices such as raising of temperatures may affect the variability in larval development. Overall, the presented modeling approach provides a support system for decisions that must be made for the timely delivery of larvae and reduction of variability.

Keywords: insect rearing; dynamic energy budget (DEB) theory; *Spodoptera frugiperda*; development; temperature; variability

Citation: Gergs, A.; Baden, C.U. A Dynamic Energy Budget Approach for the Prediction of Development Times and Variability in *Spodoptera frugiperda* Rearing. *Insects* **2021**, *12*, 300. https://doi.org/10.3390/insects12040300

Academic Editor: Gaelle Le Goff

Received: 23 February 2021
Accepted: 25 March 2021
Published: 29 March 2021

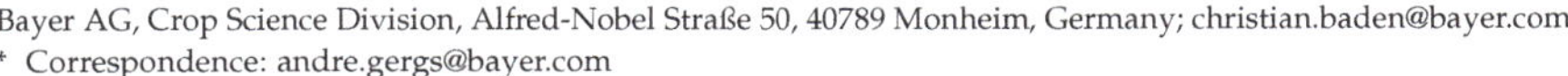

1. Introduction

Laboratory experiments are usually designed in a way that allows the study of biological behaviors in response to the variation of one factor only, while keeping the others as constant as possible. Attempts are made to reduce the variability in the initial conditions of the experiments to increase their statistical power. As such, the experimental outcome is directly linked to the rearing conditions of the biological test organisms and their homogenous state.

Due to its widespread and expanding dispersal and its economic and socioeconomic relevance [1], the fall armyworm, *Spodoptera frugiperda* (J.E. Smith) (Lepidoptera, Noctuidae), is an increasingly important laboratory and field model organism in biological and agricultural research. The fall armyworm is native to the Americas, but in the last few years, it has expanded its distribution over large parts of Africa, Asia, and Australia [2] This polyphagous noctuid species causes major economic damage, as the 350 known host plants species includes many important crops as well, e.g., corn, rice, sorghum, cotton, and many varieties of vegetables [3]. In addition to its polyphagous abilities, the fall armyworm has other properties that make it a severe pest. These include the absence of a diapause, a high reproductive rate, and high migratory abilities [4–6].

The artificial rearing of this important pest species is crucial to obtain more information about its biology, behavior, metabolic rate, and all the other puzzle pieces needed for an integrated pest management program [7]. This becomes even more urgent because of the expanding distribution range of *S. frugiperda*; therefore, rearing conditions and artificial diets are the topics of numerous current publications [7–9]. The important influence of the temperature regime on the development times of different larval stages was recently shown by Du Plessis et al. [8], as well as Lopez et al. [10] and Montezano et al. [3].

In this paper, we illustrate how resource availability and ambient temperature may affect larval development and variability by means of dynamic energy budget (DEB) modeling. DEB theory is based on the idea that the principle of energy metabolism is largely conserved across species. DEB models can, thus, be applied to simulate the entire life cycle of a wide range of animals [11], including insects [12–14]. First proposed by Kooijman [15], DEB theory has previously been applied to analyze the effects of environmental factors on life history processes [16,17], including temperature [18] and food availability in a population context [19,20].

In this paper, we aimed to analyze how resource availability in laboratory cultures and rearing practices, such as growth-controlling temperatures, affect larval development and variability in *S. frugiperda*.

2. Materials and Methods

DEB models quantify the rates at which organisms assimilate energy from the environment and subsequently allocate this energy via a reserve compartment to structural growth, the reproductive system, and the maintenance of bodily functions. The models describe the entire life cycle of an organism starting with the embryonic phase to quantify somatic (structural) growth, maturation, and reproduction. Maturity thresholds for birth and puberty mark the onset of feeding and investment of energy into reproduction, respectively. As all modeled rates are dependent on temperature and energy acquisition as a function of food availability, both environmental factors are key drivers in DEB model systems. Many decades of research in DEB theory have led to the development of the Add-my-Pet (AmP) database, where data, models, and sets of parameters for more than 2800 species have currently been collected. Within the collection, the parameter set for a given species is available and can be used in combination with the well-tested model code that is used for the parameter estimation. All of the data used for estimation, as well as the assumptions used to link data and the model, are well documented and openly accessible. Herein, we used a DEB model for *S. frugiperda* previously been developed and published via the AmP collection [21]. The model makes use of the "hex" variant of the DEB system to cover the life cycles of holometabolic insects and some other hexapods. The model covers the morphological life stages of the egg, larva, pupa, and the imago, as well as the functional stages, including the embryo and reproductive stages. As a deviation from the standard model [22], the larval stage accelerates in terms of growth [23] and allocates energy to reproduction. Larval stage transitions are marked by instar-specific stresses, depending on their structural length. For further details of the model implementation and parameters see [21]. Exemplified code is available from the Supplementary Material S2.

For all scenario analyses and the simulation of developmental variability, we included stochasticity in the model: We multiplied the maximum specific assimilation flux (p_{Am}) which is one of the primary DEB parameters, with a random factor drawn from a normal distribution with a mean of one. The range of variability for a given setting was explored by means of Monte-Carlo simulations, i.e., repeated simulation runs for different random factors. Note that by posing variability on the parameter p_{Am} and employing a constant value for the scaled functional response f, we affectively assumed that the assimilation efficiency differs among individual larvae.

For model validation, we mimicked the variable laboratory conditions as employed in an independent experiment and compared the model output and data. The culture conditions and experimental set-up were as follows. In *S. frugiperda* rearing, we used cylindrical flight cages (diameter, 24 cm; height, 26 cm) containing 200 adults (male/female ratio, approximately 1:1) fed with a sugar solution. The temperature was 25 °C, humidity 50%, and the flight cages were transparent, allowing exposure to a dimmed natural light regime. Females deposited their eggs in batches on a piece of paper kitchen towel, forming the lid of the cylindrical flight cages. Eggs were allowed to mature and hatch at 26 °C and in darkness in plastic tubs (18 cm length × 13.5 cm width × 6 cm height) containing a layer of artificial diet. At the second larval stage, the larvae were separated into individual Petri dishes (diameter, 5.5 cm) and cultured at 25 °C, 60% humidity, and in darkness until pupation. The recipe for 1 L of the artificial diet was: 0.43 L of hot water to solve 14.5 g of agar and 18 g of alfalfa powder, and then after adding 0.325 L of cold water, the other ingredients, 0.5 mL of rapeseed oil, 40 g of baker's yeast, 0.7 g of Wesson Salt Mix, 1 g of β-Sistosterol, 0.5 g of L-Leucin, 4 g of ascorbic acid, 3.7 g of vitamin mix, 3.6 g of sorbic acid, 133 g of bean flour, 0.5 g of antibiotics, and 3 mL of 4- hydroxybenzene S were added. The validation experiment was performed using a variable-temperature regime and was started with one of the mass rearing plastic containers derived from the laboratory culture. To prevent cannibalism [24], 36 larvae were separated during the second larval instar into small Petri dishes (5.5 cm in diameter) filled with an ad libitum amount of artificial diet. These Petri dishes were shifted between different temperature regimes using a series of climate cabinets maintained at 22, 26, and 30 °C. The larval stage attained was determined individually once to twice a day.

3. Results

We used a DEB model for *S. frugiperda* to simulate larval development times under different rearing conditions. The simulated larval development times, on average, ranged from 9.0 to 31.2 days at 32 °C and 18 °C, respectively (Figure 1), assuming ad libitum food conditions. In order to account for the variability in development times that were usually observed in laboratory experiments, we estimated a standard deviation in the maximum specific assimilation flux, one of the primary DEB parameters, of 0.087 J d^{-1} cm^{-2}, based on the empirical range observed by Du Plessis et al. [8]. The resulting variability in larval development times is shown in Figure 1.

For validation, model predictions were compared to experimental data (Supplementary Material S1) that had not been used for model development. In this laboratory experiment, ambient temperature was varied between 22 and 30 °C. Starting from the first instar, under these conditions, all larvae reached the sixth instar within 16 days. The DEB model accurately predicted the larval stage transitions over time—most of the measured data points were within the prediction interval (Figure 2), indicative of a good model performance. Interestingly, the modeled variability increased within the fourth instar, coinciding with a drop in ambient temperature at day 10 of the experiment.

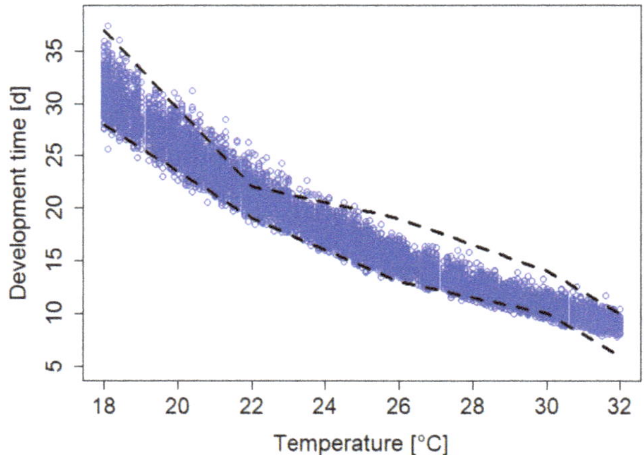

Figure 1. Simulated (dots) and measured range (dashed lines; Du Plessis et al. [8]) of the individual larval development times in days (d) in *S. frugiperda*.

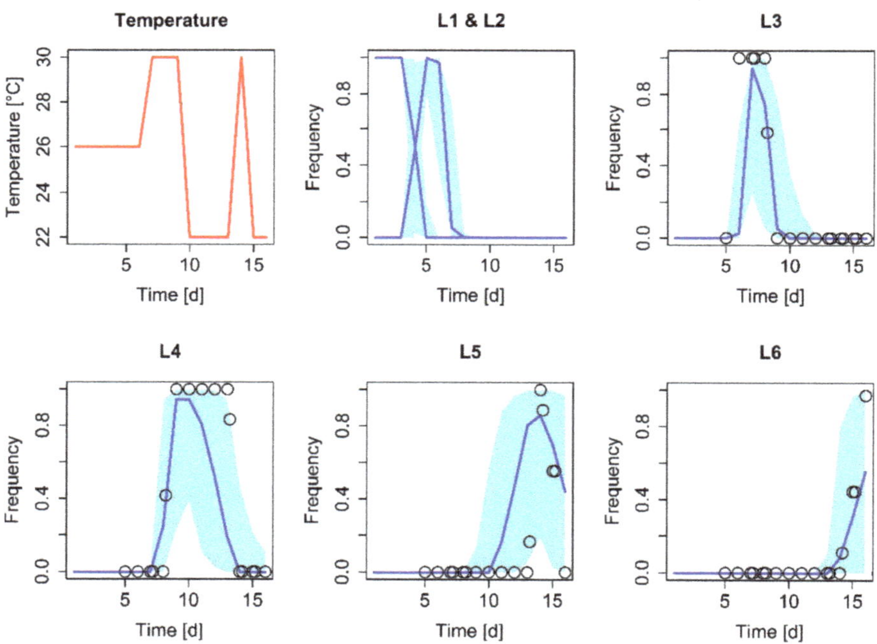

Figure 2. Frequency of the larval instars measured over time in days (d) for an exemplified variable temperature scenario (**red line**). The median and 95% confidence limits of the model prediction are represented by the blue lines and shade areas, respectively. Dots indicate measured data. The instar frequencies were recorded twice per day from instar 3.

In the scenario analyses, we subsequently analyzed the variability in larval development times with respect to food availability and fluctuating temperatures in more detail. We varied the value for the scaled functional response in the DEB model to simulate different experimental food conditions and observed the model outputs in terms of development times, starting from the first instar larvae. The scaled functional response can have values between f = 1 (ad libitum) and f = 0 (absence of food) and might represent different food qualities or quantities. The model simulations revealed that variability in larval development times may increase with decreasing food quality or quantity, as illustrated for the fifth instar larvae in Figure 3. Moreover, the simulated development times increased with increasing values for the scaled functional response, as did the standard error (Figure 5).

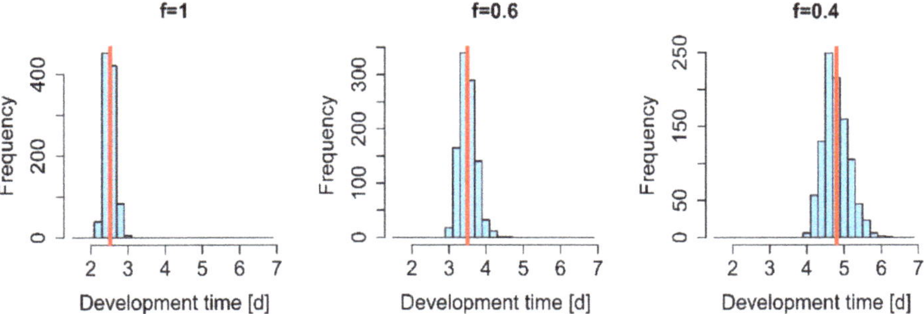

Figure 3. Simulated development times in days (d) (n = 1000) for the example of the fifth larval instar reared at 26 °C and three different feeding regimes: Ad libitum (f = 1), intermediate (f = 0.6), and low (f = 0.4) food availability. Blue columns represent the frequencies of the development times of larvae (n = 1000), and the red vertical line indicates the mean development time per instar.

The simulated larval development times were also affected by up-/downregulation of ambient temperatures (Figure 4). We decreased the simulated temperatures between 30 and 22 °C just before the transition into the third larval instar, and vice versa. This model analysis revealed that the downregulation of ambient temperatures may result in increased variability (Figure 4, left), while the opposite was the case for increasing temperature regimes (Figure 4, right).

4. Discussion

The DEB model provides an excellent basis for the prediction of the developmental behavior in *S. frugiperda* rearing, particularly regarding changes in temperature and provision of nutrients. The presented model adequately predicts development times in a changing rearing environment and allows for the analysis of variability among individuals.

Increased variability, with longer development times at lower ambient temperatures, were reported for *S. frugiperda* by Du Plessis et al. [8] (Figure 1) and Garcia et al. [25]. The authors of these studies fitted linear regression models to temperature data for different stages to evaluate the development times and to estimate the temperature thresholds and the number of degree-days. However, Du Plessis et al. [8] argued that the developmental rates become non-linear at unfavorable temperatures. Deviations of the linear temperature model were thus employed, e.g., by Garcia et al. [26], to describe the development rates in *S. frugiperda* beyond empirical observations.

DEB models, by contrast, are based on first principles of bioenergetics. They capture basic life history processes, such as feeding, development, growth, maintenance of bodily functions, reproduction, and senescence under environmental fluctuations in temperature and food availability in one coherent framework and a relatively low number of parameters [11]. The core concept of DEB theory is consistent with the principles of thermodynamics [27,28] and general trends in evolutionary history [29]. The modular structure of DEB models allows specific environmental attributes or stressors to be accounted for without changing the core of the theory. In DEB models, changes in life history trajectories, such as growth and development in response to environmental temperatures, are considered to be a result of changes in bioenergetic rate constants. It has been demonstrated by means of DEB modeling that different life history processes, such as food ingestion, growth, and reproduction, share the same Arrhenius temperature [18].

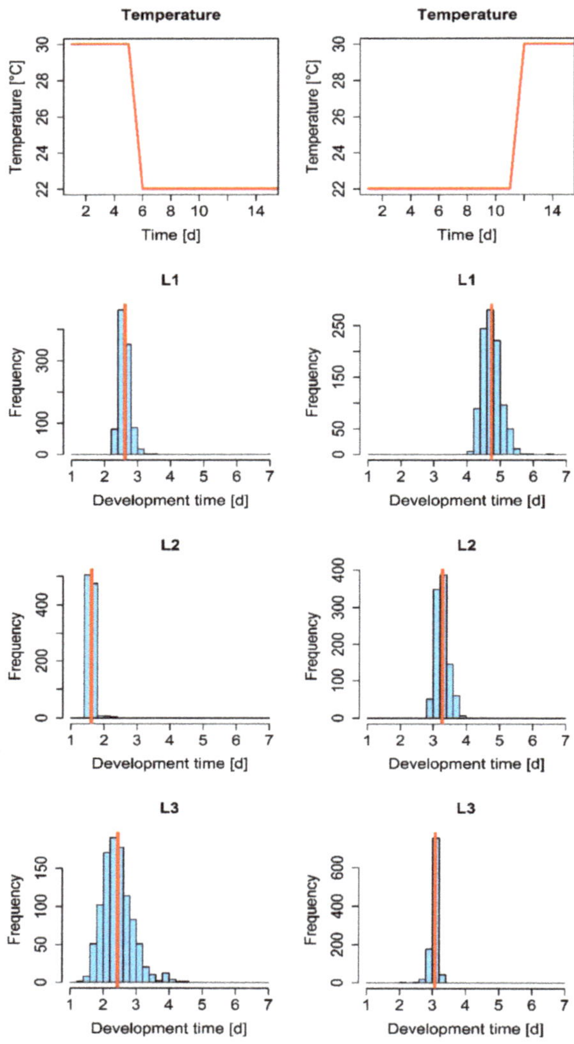

Figure 4. Simulated larval development for decelerating (**left**) and accelerating (**right**) temperature regimes. Blue columns represent the frequencies of the development times of larvae (n = 1000), and the red vertical line in the histograms indicates the mean development time per instar.

In the DEB model, the three-parameter Arrhenius model that was parameterized for *S. frugiperda* [21] describes the non-linear temperature response in larval development times well (Figure 1). Moreover, the same Arrhenius parameter described the instar-specific developmental times well, except for the first instar, where a deviating parameter value was needed [21]. The DEB model approach thus requires fewer parameters and less experimental testing compared to stage-specific linear regression approaches. As regards food, the modeled increase in variability with decreasing values for the scaled functional response f (Figure 3) captures the general pattern of standard errors being increased if development times are prolonged—see, e.g., daSilva et al. [30] for different food types such as oats, cotton, or an artificial diet (Figure 5). Note that simulated development times represent the time span from egg hatching until pupation, while the data [30] also cover the pupa stage, which explains the difference in the development times and the standard errors between the simulation and data. No extra parameters were needed to simulate the different food levels, as the energy acquisition from the environment and subsequent bioenergetic fluxes are inherent properties of the DEB model.

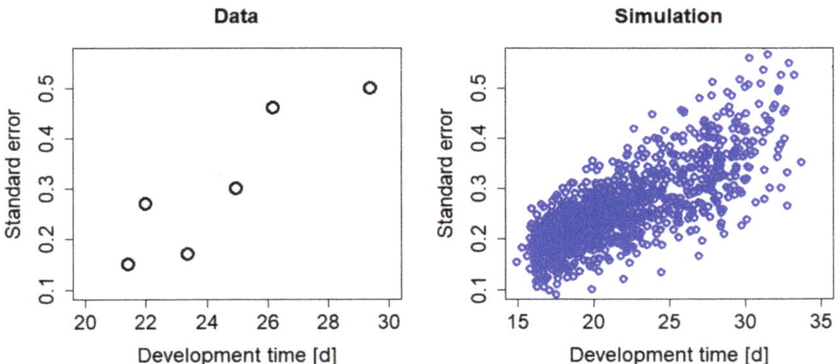

Figure 5. Standard error as function of development time in days (d). The data cover development times (larvae to adults), as measured by daSilva et al. [30], for different food types (soybean, cotton, maize, oats, wheat, and an artificial diet) at an ambient temperature of 25 ± 2 °C. Simulated mean development times (first instar to pupation) and standard errors (n = 10) were calculated 1000 times for random values of the scaled functional response between f = 0.4 and f = 1 and temperatures in the range of 25 ± 2 °C.

The DEB model revealed that lower temperatures and food sources of low nutritional value lead to more heterogenous development times (Figures 3 and 4), which subsequently result in more heterogenous colonies or cohorts (e.g., different larval stages). We gained these results both via the model and our experimental data independently of one another (Figure 4). As these effects lead to more homogenous or heterogenous colonies even within a rather short time (Figure 4), every change in a temperature regime should be conducted thoughtfully. For example, the lowering of temperatures overnight (e.g., for slowing down the growth rate to ensure the desired larval stage for a given experiment day) would lead to a higher developmental diversity within the cohort, whereas a higher temperature would have the opposite effect. Therefore, for more homogenous rearing, conditions should include a rather high temperature, between 26 and 30 °C [8], with a good quality food source (Figure 3) [7] to ensure reproducible experimental data with less variability in *S. frugiperda* research.

If the temperature conditions should change during rearing, the DEB model for *S. frugiperda* is a useful tool to predict and understand the resulting changes in development times and variability. On the contrary, with the help of the model, rearing conditions can be calculated and predicted beforehand, decreasing the need to interfere in running larval rearing. The DEB model can therefore enable researchers to obtain the desired larval stage at a specific date.

A further potential application of the model is for estimating the development rates in field populations of *S. frugiperda*, providing an alternative calculation type besides the degree-days model [8], if there is sufficient information about the temperature and host plant nutrients available (e.g., *S. frugiperda* larval development times on cotton are much longer than on maize [31]). Therefore, DEB models provide important tools for future insect pest management research in laboratory (rearing) and field population biology. They could enhance risk management tools, e.g., for seasonal multigenerational migration, to further improve existing population dynamics and migration models [25,26]. For instance, the NicheMapR package can be used to make field calculations as a function of microclimate using the DEB model [32].

5. Conclusions

The dynamic energy budget model is a powerful tool to predict the development times in *S. frugiperda* rearing and can provide guidance on how to adjust the development times to individual needs. Furthermore, it helps to reduce possible side effects by avoiding unnecessary developmental variabilities within a larval cohort caused by unfavorable temperature or nutritional regimes. Many potential interactions with other models predicting population dynamics in the field can be envisaged.

Supplementary Materials: The following are available online at https://www.mdpi.com/article/10.3390/insects12040300/s1: S1: Validation data and S2: model code.

Author Contributions: Conceptualization, A.G. and C.U.B.; methodology, A.G. and C.U.B.; software, A.G.; validation, A.G. and C.U.B.; formal analysis, A.G.; investigation, A.G. and C.U.B.; data curation, A.G.; writing—original draft preparation, A.G. and C.U.B.; writing—review and editing, A.G. and C.U.B.; visualization, A.G. All authors read and agreed to the published version of the manuscript.

Funding: This research received no external funding.

Data Availability Statement: Validation data (Figure 2) are available in the Supplementary Materials.

Acknowledgments: The authors thank Petra Papachristou, Dorina Pöttgen, Tina Teiwes, and Markus Schaeben for the maintenance of the *Spodoptera frugiperda* rearing and their technical support.

Conflicts of Interest: The authors are employed by Bayer AG. Bayer products were used in the fall armyworm control. The authors declare no additional conflicts of interest.

References

1. EPPO. Available online: https://gd.eppo.int/taxon/LAPHFR/distribution (accessed on 8 December 2020).
2. FAO. The Global Action for Fall Armyworm Control: Action framework 2020–2022. In *Working Together to Tame the Global Threat*; Food and Agriculture Organization of the United Nations: Rome, Italy, 2020. [CrossRef]
3. Montezano, D.G.; Specht, A.; Sosa-Gómez, D.R.; Roque-Specht, V.F.; Sousa-silva, J.C.; Paula-Moraes, S.V.; Peterson, J.A.; Hunt, T.E. Host plants of *Spodoptera frugiperda* (Lepidoptera: Noctuidae) in the Americas. *Afr. Entomol.* **2018**, *26*, 286–300. [CrossRef]
4. Pair, S.D.; Raulston, J.R.; Sparks, A.N.; Westbrook, J.K.; Douce, G.K. Fall armyworm distribution and population dynamics in the southeastern states. *Fla. Entomol.* **1986**, *69*, 468–487. [CrossRef]
5. Johnson, S.J. Migration and the life history strategy of the fall armyworm, *Spodoptera frugiperda* in the western hemisphere. *Int. J. Trop. Insect Sci.* **1987**, *8*, 543–549. [CrossRef]
6. Westbrook, J.; Fleischer, S.; Jairam, S.; Meagher, R.; Nagoshi, R. Multigenerational migration of fall armyworm, a pest insect. *Ecosphere* **2019**, *10*, 1–12. [CrossRef]
7. Pinto, J.R.L.; Torres, A.F.; Truzi, C.C.; Vieira, N.F.; Vacari, A.M.; De Bortoli, S.A. Artificial Corn-Based Diet for Rearing *Spodoptera frugiperda* (Lepidoptera: Noctuidae). *J. Insect Sci.* **2019**, *19*, 1–8. [CrossRef]
8. Du Plessis, H.; Schlemmer, M.-L.; Van den Berg, J. The Effect of Temperature on the Development of *Spodoptera frugiperda* (Lepidoptera: Noctuidae). *Insects* **2020**, *11*, 228. [CrossRef]
9. Jin, T.; Lin, Y.; Chi, H.; Xiang, K.; Ma, G.; Peng, Z.; Yi, K. Comparative Performance of the Fall Armyworm (Lepidoptera: Noctuidae) Reared on Various Cereal-Based Artificial Diets. *J. Econ. Entomol.* **2020**, *116*, 2986–2996. [CrossRef] [PubMed]
10. Lopez, R.Y.; Ortega, A.V.; Centeno, J.H.A.; Ruiz, J.S.; Carranza, J.A.Q. Sistema de alerta contra el gusano cogollero *Spodoptera frugiperda* (J.E. Smith) (Insecta: Lepidoptera: Noctuidae). *Rev. Mex. Cienc. Agríc.* **2019**, *10*, 405–416.
11. Kearney, M.R.; Domingos, T.; Nisbet, R. Dynamic Energy Budget Theory: An Efficient and General Theory for Ecology. *BioScience* **2015**, *65*, 34. [CrossRef]
12. Gergs, A.; Jager, T. Body size-mediated starvation resistance in an insect predator. *J. Anim. Ecol.* **2013**, *83*, 758–768. [CrossRef] [PubMed]
13. Maino, J.; Kearney, M. Testing mechanistic models of growth in insects. *Proc. R. Soc. B* **2015**, *282*, 1–9. [CrossRef]
14. Llandres, A.; Marques, G.; Maino, J.; Kooijman, S.A.L.M.; Kearney, M.; Casas, J. A Dynamic Energy Budget for the whole life-cycle of holometabolous insects. *Ecol. Monogr.* **2015**, *85*, 353–371. [CrossRef]
15. Kooijman, S.A.L.M. Energy budgets can explain body size relations. *J. Theor. Biol.* **1986**, *121*, 269–282. [CrossRef]
16. Pieters, B.J.; Jager, T.; Kraak, M.H.S.; Admiraal, W. Modeling responses of Daphnia magna to pesticide pulse exposure under varying food conditions: Intrinsic versus apparent sensitivity. *Ecotoxicology* **2006**, *15*, 601–608. [CrossRef] [PubMed]
17. Muller, E.B.; Nisbet, R.M. Dynamic energy budget modeling reveals the potential of future growth and calcification for the coccolithophore Emiliania huxleyi in an acidified ocean. *Glob. Chang. Biol.* **2004**, *20*, 2031–2038. [CrossRef]
18. Kooijman, S.A.L.M. *Dynamic Energy Budget Theory for Metabolic Organization*; Cambridge University Press: Cambridge, UK, 2010; p. 514.

19. Martin, B.T.; Jager, T.; Preuss, T.G.; Nisbet, R.; Grimm, V. Predicting population dynamics from the properties of individuals: A cross-level test of Dynamic Energy Budget theory. *Am. Nat.* **2013**, *181*, 506–519. [CrossRef]
20. Gergs, A.; Preuss, T.G.; Palmqvist, A. Double Trouble at High Density: Cross-Level Test of Resource-Related Adaptive Plasticity and Crowding-Related Fitness. *PLoS ONE* **2014**, *9*, e91503. [CrossRef] [PubMed]
21. Gergs, A.; Baden, C.U. AmP Spodoptera frugiperda, version 2020/07/21. 2020. Available online: https://www.bio.vu.nl/thb/deb/deblab/add_my_pet/entries_web/Spodoptera_frugiperda/Spodoptera_frugiperda_res.html (accessed on 8 December 2020).
22. Kooijman, S.A.L.M.; Sousa, T.; Pecquerie, L.; Van der Meer, J.; Jager, T. From food dependent statistics to metabolic parameters, a practical guide to the use of dynamic energy budget theory. *Biol. Rev.* **2008**, *83*, 533–552. [CrossRef] [PubMed]
23. Kooijman, S.A.L.M. Metabolic acceleration in animal ontogeny: An evolutionary perspective. *J. Sea Res.* **2014**, *94*, 128–137. [CrossRef]
24. Chapman, J.W.; Williams, T.; Escribano, A.; Caballero, P.; Cave, R.D.; Goulson, D. Fitness consequences of cannibalism in the fall armyworm, *Spodoptera frugiperda*. *Behav. Ecol.* **1999**, *10*, 298–303. [CrossRef]
25. Garcia, A.; Godoy, W.A.C.; Thomas, J.M.C.; Nagoshi, R.N.; Meagher, R. Delimiting Strategic Zones for the Development of Fall Armyworm (Lepidoptera: Noctuidae) on Corn in the State of Florida. *J. Econ. Entomol.* **2017**, *111*, 120–126. [CrossRef] [PubMed]
26. Garcia, A.G.; Ferreira, C.P.; Godoy, W.A.C.; Meagher, R. A computational model to predict the population dynamics of *Spodoptera frugiperda*. *J. Pest. Sci.* **2019**, *92*, 429–441. [CrossRef]
27. Jusup, M.; Sousa, T.; Domingos, T.; Labinac, V.; Marn, N.; Wang, Z.; Klanjscek, T. Physics of metabolic organization. *Phys. Life Rev.* **2017**, *20*, 1–39. [CrossRef] [PubMed]
28. Sousa, T.; Domingos, T.; Poggiale, J.C.; Kooijman, S. Dynamic energy budget theory restores coherence in biology Introduction. *Philos. Trans. R. Soc. Lond. B Biol. Sci.* **2010**, *365*, 3413–3428. [CrossRef] [PubMed]
29. Kooijman, S.A.L.M.; Troost, T.A. Quantitative steps in the evolution of metabolic organisation as specified by the Dynamic Energy Budget theory. *Biol. Rev.* **2007**, *82*, 113–142. [CrossRef] [PubMed]
30. Da Silva, D.M.; de Freitas Bueno, A.; Andrade, K.; dos Santos Stecca, C.; Oliveira, P.M.; Neves, J.; de Oliveira, M.C.N. Biology and nutrition of *Spodoptera frugiperda* (Lepidoptera: Noctuidae) fed on different food sources. *Sci. Agric.* **2017**, *74*, 18–31. [CrossRef]
31. Ali, A.; Luttrell, R.G.; Schneider, J.C. Effects of temperature and larval diet on development of the fall armyworm (Lepidoptera: Noctuidae). *Ann. Entomol. Soc. Am.* **1990**, *83*, 725–733. [CrossRef]
32. Kearney, M.R.; Porter, W.P. NicheMapR—An R package for biophysical modelling: The ectotherm and Dynamic Energy Budget models. *Ecography* **2020**, *43*, 85–96. [CrossRef]

 *insects*

Article

Geographic Monitoring of Insecticide Resistance Mutations in Native and Invasive Populations of the Fall Armyworm

Sudeeptha Yainna [1,2], Nicolas Nègre [1], Pierre J. Silvie [2,3,4], Thierry Brévault [3,4], Wee Tek Tay [5], Karl Gordon [5], Emmanuelle dAlençon [1], Thomas Walsh [5,*] and Kiwoong Nam [1,*]

1   DGIMI, University of Montpellier, INRAE, F-34095 Montpellier, France; sudeepha.yainna-kumarihami@inrae.fr (S.Y.); nicolas.negre@umontpellier.fr (N.N.); emmanuelle.d-alencon@inrae.fr (E.d.)
2   CIRAD, UPR AIDA, F-34398 Montpellier, France; pierre.silvie@cirad.fr
3   PHIM Plant Health Institute, University of Montpellier, IRD, CIRAD, INRAE, Institut Agro, F-34398 Montpellier, France; thierry.brevault@cirad.fr
4   AIDA, University of Montpellier, CIRAD, F-34398 Montpellier, France
5   Black Mountain Laboratories, CSIRO, Canberra, ACT 2601, Australia; weetek.tay@csiro.au (W.T.T.); karl.gordon@csiro.au (K.G.)
*   Correspondence: tom.walsh@csiro.au (T.W.); ki-woong.nam@inrae.fr (K.N.)

Simple Summary: The moth fall armyworm (*Spodoptera frugiperda*) is a major agricultural pest insect damaging a wide range of crops, especially corn. Field evolved resistance against *Bacillus thuringiensis* (Bt) toxins and synthetic insecticides has been repeatedly reported. While the fall armyworm is native to the Americas, its biological invasion was first reported from West Africa in 2016. Since then, this pest has been detected across sub-Saharan and North Africa, Asia, and Oceania. Here, we examine the geographical distribution of mutations causing resistance against Bt or synthetic insecticides to test if the invasion was accompanied by the spread of resistance mutations using 177 individuals collected from 12 geographic populations including North and South America, West and East Africa, India, and China. We observed that Bt resistance mutations generated in Puerto Rico or Brazil were found only from their native populations, while invasive populations had higher copy numbers of cytochrome P450 genes and higher proportions of resistance mutations at AChE, which are known to cause resistance against synthetic insecticides. This result explains the susceptibility to Bt insecticides and the resistance against synthetic insecticides in invasive Chinese populations. This information will be helpful in investigating the cause and consequence associated with insecticide resistance.

Abstract: Field evolved resistance to insecticides is one of the main challenges in pest control. The fall armyworm (FAW) is a lepidopteran pest species causing severe crop losses, especially corn. While native to the Americas, the presence of FAW was confirmed in West Africa in 2016. Since then, the FAW has been detected in over 70 countries covering sub-Saharan Africa, the Middle East, North Africa, South Asia, Southeast Asia, and Oceania. In this study, we tested whether this invasion was accompanied by the spread of resistance mutations from native to invasive areas. We observed that mutations causing Bt resistance at ABCC2 genes were observed only in native populations where the mutations were initially reported. Invasive populations were found to have higher gene numbers of cytochrome P450 genes than native populations and a higher proportion of multiple resistance mutations at acetylcholinesterase genes, supporting strong selective pressure for resistance against synthetic insecticides. This result explains the susceptibility to Bt insecticides and resistance to various synthetic insecticides in Chinese populations. These results highlight the necessity of regular and standardized monitoring of insecticide resistance in invasive populations using both genomic approaches and bioassay experiments.

Keywords: ABCC2; *Bacillus thuringiensis*; biological invasion; Cytochrome P450; Fall armyworm; insecticide resistance; *Spodoptera frugiperda*

 check for updates

Citation: Yainna, S.; Nègre, N.; Silvie, P.J.; Brévault, T.; Tay, W.T.; Gordon, K.; dAlençon, E.; Walsh, T.; Nam, K. Geographic Monitoring of Insecticide Resistance Mutations in Native and Invasive Populations of the Fall Armyworm. *Insects* **2021**, *12*, 468. https://doi.org/10.3390/insects 12050468

Academic Editor: Béla Darvas

Received: 31 March 2021
Accepted: 12 May 2021
Published: 18 May 2021

Publisher's Note: MDPI stays neutral with regard to jurisdictional claims in published maps and institutional affiliations.

1. Introduction

Insecticide resistance is one of the main challenges for the control of insect pests. Commonly used insecticides to control pest insects can be classified into two main types. The first group is Bt (*Bacillus thuringiensis*) toxins, which are generally produced by transgenic crops. The global areas of planted Bt crops is positively correlated with the number of Bt resistance pest species, implying that field-evolved Bt resistance is prevalent [1]. Field-evolved Bt resistance is a particularly serious issue in corn and cotton because the majority of those plants are Bt crops in the USA and because only GM-cotton expressing Bt toxins are grown in Australia. Field-evolved Bt resistance has been observed in several major insect Lepidoptera pest species including maize stalk borer (*Busseola fusca*), western corn rootworm (*Diabrotica virgifera virgifera*), cotton bollworm (*Helicoverpa zea*), Old World cotton bollworm (*Helicoverpa armigera*), *Helicoverpa punctigera*, fall armyworm (*Spodoptera frugiperda*) and pink bollworm (*Pectinophora gossypiella*) [1]. The other group is synthetic insecticides such as organophosphates, pyrethroids, neonicotinoids, organochlorides, carbamates, ryanoids, and spinosyns. By 2021, more than 600 arthropod species have been included in the arthropod pesticide resistance database (https://www.pesticideresistance.org/) (accessed on 21 January 2021).

Several genetic mechanisms of Bt resistance have been reported (reviewed in [2]). If a mutation in a midgut receptor causes reduced physical binding with Bt proteins, the toxicity of Bt can be decreased. These receptors include cadherin, aminopeptidase-N, alkaline phosphatases, and ATP-binding cassette transporters. Alternatively, altered processing of Bt prototoxin, increased immune status, sequestration of Bt protein, and accelerated recovery of epithelial integrity can also increase Bt resistance. The resistance against synthetic insecticides can be caused by detoxification genes [3] such as cytochrome P450 gene, esterase, and glutathione S transferase. In addition, mutations in calcium channels, acetylcholinesterase, and nicotinic acetylcholine receptors may also cause resistance [4–7].

Fall armyworm (FAW, *Spodoptera frugiperda*, J.E.Smith) (Lepidoptera: Noctuidae: Noctuinae) is one of the most damaging pest insects, partly due to the extreme polyphagy by feeding on at least 353 species in 76 plant families, which include several economically important crops such as corn, rice, sorghum, sugarcane, cotton, and soybean [8]. Genomic analyses have demonstrated a rapid expansion of detoxification genes [9–11], potentially associated with this extreme polyphagy by overcoming plant defense toxins from diverse plants.

Field-evolved resistance to Bt insecticide in the FAW has been reported from Puerto Rico [12,13], Brazil [14,15], Argentina [16], and the USA [17]. Five mutations causing Bt resistance have been reported. Interestingly, all these mutations were observed in the ATP Binding Cassette Subfamily C Member 2 (ABCC2) gene, which encodes ATP-binding cassette transporter proteins, implying parallel evolution. In a population from Puerto Rico, 2bp GC insertion at ABCC2 causes a frameshift mutation and a premature stop codon, which results in improper binding between ABCC2 and Bt Cry1F and Cry1A proteins [18,19]. Boaventura et al. identified three causal mutations of Cry1F resistance at ABCC2 including GY deletion, P799K/R substitution, and G1088D substitution from a Brazilian population [20]. Guan et al. also identified a 12bp insertion, which causes a frameshift mutation and a premature stop codon, from another Brazilian population [21]. As well as the Bt resistance, field-evolved resistance to synthetic insecticides has been widely reported. For example, a population in Puerto Rico was reported to have resistance against a wide range of synthetic insecticides including flubendiamide, chlorantraniliprole, methomyl, thiodicarb, permethrin, chlorpyriphos, zeta-cypermethrin, deltamethrin, triflumuron, and spinetoram [22]. We observed that P450 genes cause resistance against deltamethrin [23]. Boaventura et al. reported that I4734M and G4946E substitutions at the ryanodine receptor (RyR) cause resistance against diamides [24]. Carvalho et al. reported that resistance against organophosphate and carbamates is caused by A314S, G340A, and F402V mutations at acetylcholinesterase (AChE) and that pyrethroid resistance is caused by T929I, L932F, and L1014F mutations at the voltage-gated sodium channel (VGSC) gene [4].

FAW is native to North and South America, and its presence was first confirmed in West Africa in 2016 [25]. Since then, FAW has been detected across almost the entire Sub-Saharan Africa, in northern Africa/the Near East (Egypt), Middle East Asia, South Asia, South East Asia, East Asia, Australia, and more recently, the Canary Islands located off the coast between Western Africa and Morocco, and New Caledonia in the South Pacific (https://www.cabi.org/ISC/fallarmyworm) (accessed on 21 January 2021). The invasion caused a serious reduction in corn production, between 21–53% in Africa [26]. While native populations are composed of two sympatric strains, corn (sfC) and rice strains (sfR), named after their preferred host-plants [27–29], the invasive populations are observed as hybrids between corn and rice strains [30–32], but are predominantly observed in corn fields. Genomic analyses support the multiple introductions of FAW from multiple native populations including Florida and Brazil to multiple invasive areas including China and Africa [31,33]. However, the invasive populations exhibited much more homogeneous DNA sequences than native populations, implying that these invasive populations had experienced extensive admixture since their establishment across the Old World [31].

The invasion process opens up the possibility that resistance mutations have been spread as well in the invasive area. Recently, Boaventura et al. observed that resistance mutations at AChE were frequently observed from two invasive populations in Kenya and Indonesia and two populations in Brazil and Puerto Rico, while GY deletion of ABCC2 genes was observed only in the Brazilian population [34]. Boaventura et al. also reported the presence of the VGSC L1014F allele with low frequency in an Indonesian population that has, to-date, yet to be reported in other invasive populations [34]. Zhang et al. reported that the 2 bp GC insertion was not observed from Chinese populations [32], while Guan et al. reported that the 12 bp insertion in exon 15 of the ABCC2 gene was not detected in FAW populations from Uganda, Malawi, and from six Chinese provinces [21]. However, we still lack a comprehensive understanding of the geographic distribution of these resistance mutations because global profiling of the causal resistance mutations has not yet been performed. We performed whole-genome resequencing from 177 individuals from both invasive and native populations sampled from 12 geographic populations in North America, South America, Africa, India, and China to investigate adaptive evolution associated with the invasive success in our previous study [31]. In this study, we undertook reanalysis of the resequencing data of the FAW populations to test the presence of causal resistance mutations against insecticides in the FAWs.

2. Methods

2.1. Resequencing Data

Resequencing data generated in our previous study were reused [31,33]. More specifically, this dataset included 29 sfC and 49 sfR individuals from invasive populations (Benin, Malawi, Uganda, China, and India) and 70 sfC and 29 sfR individuals from native populations (Mississippi, Florida, Puerto Rico, Brazil, French Guinea, Guadeloupe, and Mexico) (Figure 1) (please see Table S1 for more information). This dataset includes four Bt-resistant and six susceptible Brazilian individuals. Paired-end whole genome resequencing was performed with approximately 20X coverage with 300 bp insert size, and 150 bp read length using Illumina (San Diego, CA, USA) HiSeq 2500, HiSeq 4000, and Novaseq 6000. Variant calling was performed using the GATK-4.0.11.0 package [35] for single nucleotide variant (SNV) and CNVCaller [36] for copy number variation (CNV). The numbers of variations were 27,117,672 and 22,916 for SNV and CNV, respectively.

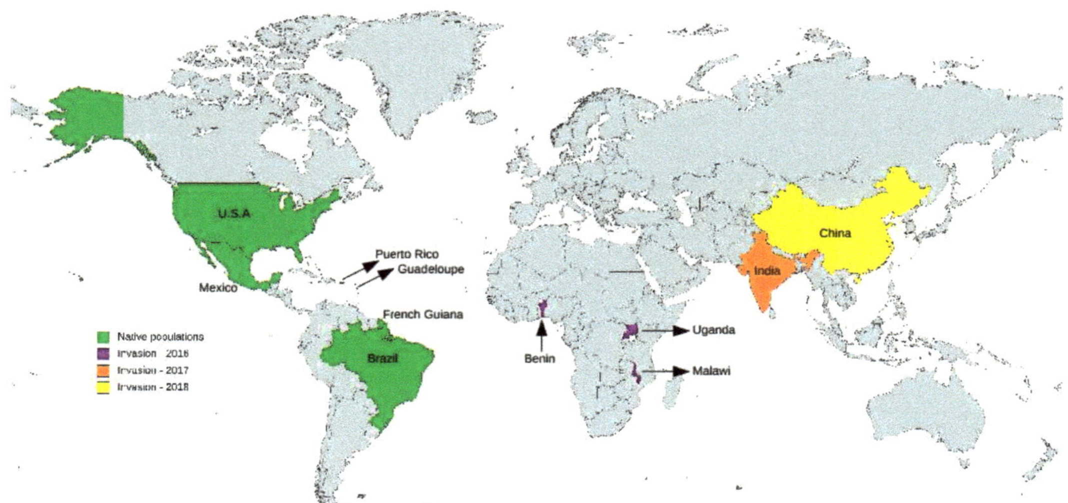

Figure 1. Countries where the analyzed individuals were sampled. The green color indicates native populations. Purple (2016), orange (2017), and yellow colors (2018) indicate the reported years of detection in countries from where individuals were collected. The map was generated using mapchart [37].

2.2. Gene Annotation and Statistical Analysis

ABCC2, RyR, AChE, and VGSC were annotated by aligning protein sequences ob tained from NCBI and reference genomes using exonerate 2.2.0 [38] with the protein2genome model. The accession numbers of these genes were QGS83596, XP_022819835, MK226188 and KC435025, for ABCC2, AChE, RyR, and VGSC, respectively. Genetic variations within these genes were identified from the resequencing data [31] using IGV tools [39]. P450 genes were identified from the gene annotation files generated in our previous study (OGS7.0) [31]. The identification of the P450 clan was performed by blasting against OGS2.2 [9], which includes manually annotated P450 genes, with 98% cutoff using ncbi blast-2.10.0+ [40]. The Fisher's exact test was performed using the R package to test a statistical significance of increased gene duplication of P450 genes. Weir and Cockerham's F_{ST} [41] between Bt susceptible and resistant individuals was calculated at the ABCC2 gene using vcftools v0.1.13 [42]. Increased F_{ST} at ABCC2 was tested by calculating the proportion of F_{ST} at randomly chosen loci with the same length of ABCC2 from whole genome sequences with 100,000 replications.

3. Results

3.1. Bt Insecticide Resistance–ABCC2 Gene

We tested the presence of five reported mutations at the ABCC2 gene causing the resistance against Bt insecticide. These mutations include (i) 2 bp GC insertion leading to a premature stop codon due to the frameshift (originally identified from a population in Puerto Rico) [18,19]; (ii) GY deletion (Brazil) [20]; (iii) 12 bp insertion causing premature stop codon due to frameshift (Brazil) [21]; (iv) P799K/R substitution (Brazil) [20]; and (v) G1088D substitution (Brazil) [20] (Figure 2). We identified the 2 bp GC insertion from 11 individuals from Puerto Rico in the resequencing data (Table 1). As this mutation was originally reported from a population in Puerto Rico, this result supports that this mutation remained in this original population in Puerto Rico and did not spread to the Old World by invasion.

1 MMDKSNKNTASNGTAVGEPKERVRKKPNILSRIFVWWIFPVLITGNKRDVEEDDLIVPSKKFNSERQGEYFERYWFEEVA80

81IAEREDRDPSLWKAMRRAYWLQYMPGAIFVLLISGLRTAQPLLFSQLLSYWSVDSEMSQQDAGLYALAMLGINFITMMCT160

161HHNNLFVMRFSMKVKIAASSLLFRKLLRMSQVSVGDVAGGKLVNLLSNDVARFDYAFMFLHYLWVVPLQVGVVLYFVYDA240

241AGWAPYVGLFGVIILIMPLQAGLTKLTGVVRRMTAKRTDKRIKLMSEIINGIQVIKMYAWEKPFQLVVKAARAYEMSALRKSI323

324FIRSMFLGFMLFTERSVMFLTVLTLALTGNMISATLIYPIQQYFGIITMNVTLILPMAFASFSEMLISLERIQGFLLLDERSDIQI409

410TPKVVNGAGSKLFNNSKKEGGLETGIVLPTKYSPTEANMARPMQDEPNMADYPVQLNKVNATWADLNDNKEMTLKNISL488

489RVRKNKLCAVIGPVGSGKTSLLQLLLRELPVTSGNLSISGTVSYASQEPWLFPATVRENILFGLEYNVAKYKEVCKVCSLLP570

571DFKQFPYGDLSLVGERGVSLSGGQRARINLARAVYREADIYLLDDPLSAVDANVGRQLFDGCIKGYLSGKTCILVTHQIHYL652

653KAADFIVVLNEGSVENMGSYDELMKTGTEFSMLLSDQASEGSDTDKKERPAMMRGISKMSVKSDDEEGEEKVQVLEAEE731

732RQSGSLKWDVLGRYMKSVNSWCMVVMAFLVLVITQGAATTTDYWLSFWTNQVDGYIQTLPEGESPNPELNTQVGLLTTG810

811QYLIVHGSVVLAIIILTQVRILSFVVMTMRASENLHNTIYEKLIVAVMRFFDTNPSGRVLNRFSKDMGAMDELLPRSMLETVQ893

894MYLSLASVLVLNAIALPWTLIPTTVLMFIFVFLLKWYINAAQAVKRLEGTTKSPVFGMINSTISGLSTIRSSNSQDRLLNSFDDA978

979QNLHTSAFYTFLGGSTAFGLYLDTLCLIYLGIIMSIFILGDFGELIPVGSVGLAVSQSMVLTMMLQMAAKFTADFLGQMTAVER1062

1063VLEYTKLPTEENMETGPTTPPKGWPSAGEVTFSNVYLKYSPDDPPVLKDLNFAIKSGWKVGVVGRTGAGKSSLISALFRLS1143

1144DITGSIKIDGLDTQGIAKKLLRSKISIIPQEPVLFSASLRYNLDPFDNYNDDDIWRALEQVELKESIPALDYKVSEGGTNFSMGQ1228

1229RQLVCLARAILRSNKILIMDEATANVDPQTDALIQKTIRKQFATCTVLTIAHRLNTIMDSDRVLVMDQGVAAEFDHPYILLSNPNS1314

1315KFSSMVKETGDNMSRILFEVAKTKYESDSKTA1346

GC insertion & frameshift mutation (Banerjee R et al., 2017 Sci Reports & Flagel et al., 2018 Sci Reports)

P799K/R substitution (Boaventura et al., 2020 IBMB)

GY deletion (Boaventura et al., 2020 IBMB)

12bp insertion & frameshift mutation (Guan et al., 2020, Insect Science.)

G1088D substitution (Boaventura et al., 2020 IBMB)

Figure 2. Full amino acid sequence of the ABCC2 gene with information on all mutations studied. The green and orange arrows indicate the identified and unidentified mutations from the resequencing data, respectively.

Table 1. Individuals with observed resistant mutations (* indicates individuals heterozygous for the mutations). PR: Puerto Rico, CC: susceptible Brazilian individuals, rCC: resistant Brazilian individuals.

Mutations	Individuals with Resistance Mutations
2 bp insertion and a frameshift mutation	* PR1, PR5, PR12, PR14, PR16, * PR18, * PR19, * PR27, PR30, * PR31, * PR33
GY deletion	CC44, CC69
P799K/R	CC44, CC69
12 bp insertion and a frameshift mutation	* rCC25, * rCC5

GY deletion was identified from two individuals in the Brazilian population, and P799K/R was observed from the same individuals. This result implies that this mutation also remained in the original population. Intriguingly, these mutations exist at the homozygous state in these two Bt susceptible individuals, suggesting a possibility that these mutations alone are not sufficient for Bt resistance.

We identified two individuals with 12 bp insertion only from Brazil, also supporting that this mutation remains in the original population and confirms the analysis of Guan et al. [21]. G1088D was not observed from the resequencing data.

As unidentified mutations causing Bt resistance might exist, we investigated sequences with complete genetic differentiation between resistant and susceptible Brazilian individuals. In total, 49 SNPs had complete genetic differentiation (i.e., F_{ST} = 1), and all these positions were found to be intronic. This result shows that other causal mutations are not likely to exist in the tested individuals.

We tested the possibility that natural selection on the ABCC2 gene is responsible for the field-evolved Bt resistance in the Brazilian population. In this case, ABCC2 is expected to have increased genetic differentiation between resistant and susceptible groups. F_{ST} calculated between these two groups was 0.2922 at ABCC2 gene. In total, 7.7% of 100,000 randomly chosen loci with the same length of the ABCC2 gene had F_{ST} higher than 0.2922 (equivalent to *p*-value = 0.077) (Figure 3), indicating marginally significant statistical support for increased genetic differentiation. This result implies that natural selection might have increased the Bt resistance of the Brazilian population.

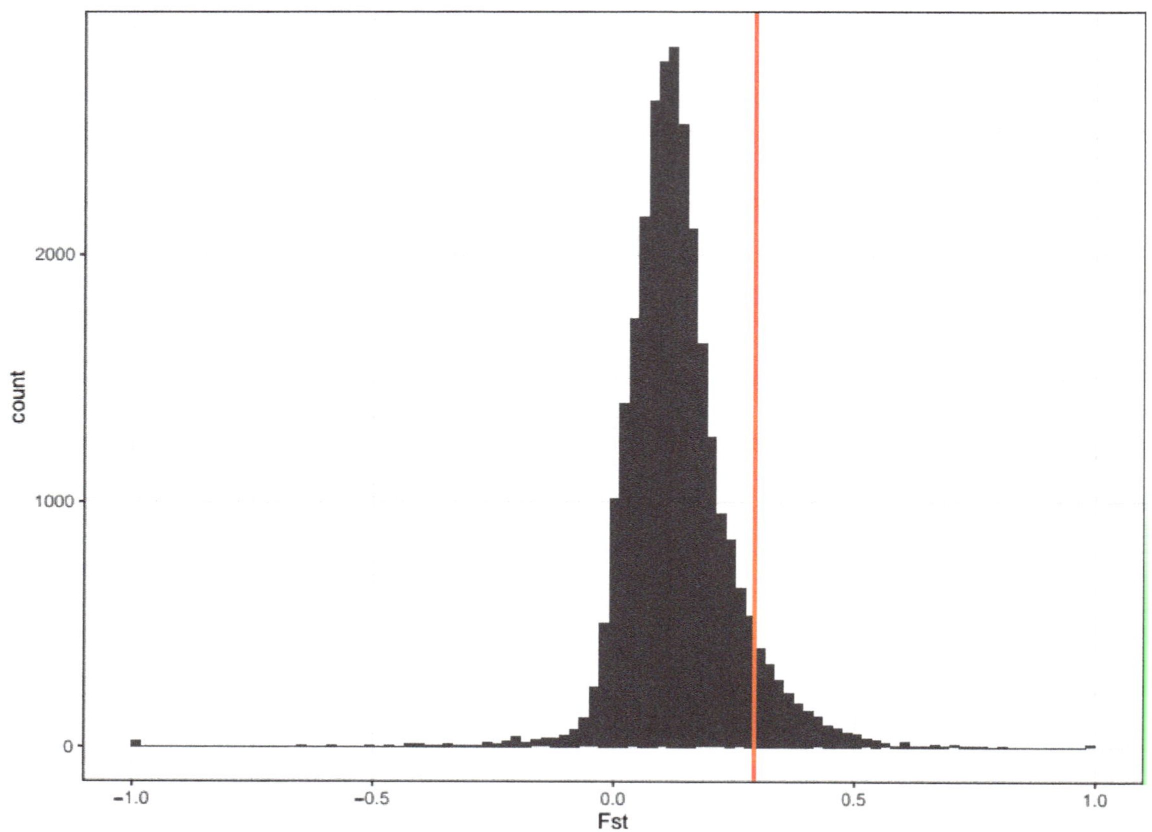

Figure 3. F_{ST} calculated between CC and rCC (red vertical line) and from random grouping among CC and rCC.

3.2. Synthetic Insecticide Resistance

P450 genes increase the level of synthetic insecticide resistance in FAW, and P450 gene duplication has been suggested as a possible cause underpinning resistance [23]. In total, 99 P450 genes were identified from the reference genome (Table S2). Copy number variation compared with the reference genome was observed from 47 P450 genes, which include4, 12, and 22 genes belonging to clan2, clan3, and clan4, respectively. In all clades, invasive populations have higher numbers of duplicated P450 genes per individual than native populations for all clades (Figure 4). On the other hand, native populations have higher numbers of deleted P450 genes per individuals than invasive populations for all clades. Consequently, invasive populations have overrepresented numbers of duplicated P450 genes compared with deleted ones than the native populations (p-value = 6.018×10^{-21}, 0.0004553, 0.04414 for clan2, clan3, and clan4, respectively; two-tailed Fisher's exact test). This result supports the hypothesis that invasive populations have higher P450 gene numbers than native populations through CNVs.

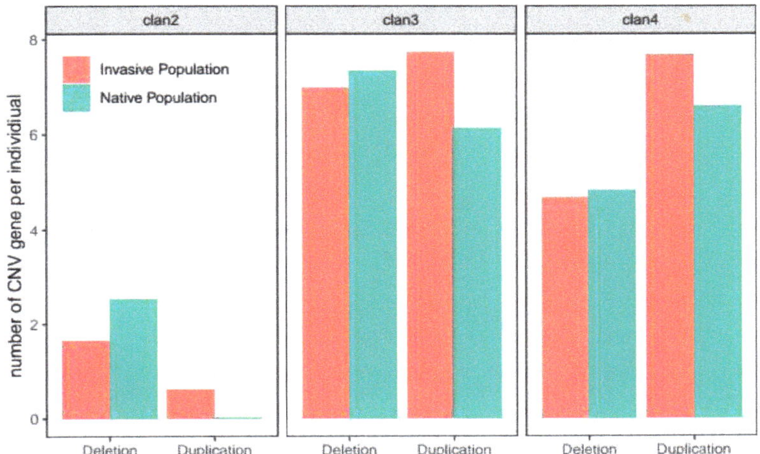

Figure 4. Average numbers of deleted or duplicated P450 genes per individual in invasive and native populations for clan2, clan3, and clan4.

The resequencing data show all three reported resistance mutations at the AChE (Figure 5), as shown by Boaventura et al. [34]. F209V mutations were observed from populations in Uganda, Malawi, Brazil, and China, as reported from Guan et al. [21] as well as India, Benin, Puerto Rico, Mexico, Mississippi, and Florida. Invasive populations have a higher proportion of individuals with resistance F209V mutations (80.26%) than native populations (72.37%) ($p = 0.001097$, two-tailed Fisher's exact test). A201S mutation was found from populations in Uganda, Malawi, and Brazil, as shown by Guan et al., but also from Benin, Mississippi, and Florida. Invasive populations have a higher proportion of individuals with A201S resistance mutations (20.0%) than native populations (7.14%) ($p = 0.01966$). Interestingly, two codons were found for the susceptible 'A' in the population from Benin (GCG and GCA). In Benin, the GCG codon was observed from 28 individuals (as well as other populations). The GCA codon was observed from three individuals in Benin. The most parsimonious explanation is that the GCA codon was newly generated from GCG by a point mutation at the third codon position. The G227A mutation was found in Brazil, as per Guan et al. [21] as well as in Puerto Rico and Florida. The proportion of individuals with this resistance mutation was 17.65% in native populations. Resistance mutations at RyR (I4734M and G4946E substitutions) or the VGSC (T929I, L932F, and L1014F substitutions) were not observed in the analyzed populations of FAW.

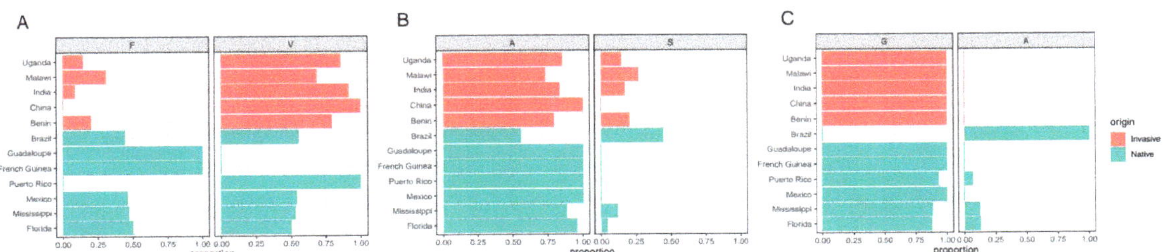

Figure 5. The proportion of individuals with susceptibility and resistance at AChE for (**A**) F209V, (**B**) A201S, and (**C**) G227A in invasive and native populations.

4. Discussion

The global spread of FAW through invasion may be accompanied by the spread of resistance mutations to Bt or synthetic insecticides to invasive areas. This is known to occur in other species. The population of the cotton bollworm *Helicoverpa armigera* is thought to have invaded South America with a pyrethroid resistance allele, which was then shared with a sister species, *H. zea* [43,44]. Therefore, monitoring the geographical distribution of resistance mutations in the context of invasion can provide useful information for pest FAW control and insecticide resistance management. We identified known resistance mutations from the whole genome sequences of 177 individuals collected from 12 geographic populations in North America, South America, Africa, India, and China. We showed that resistance Bt mutations at ABCC2 remained at their origin in Puerto Rico and Brazil, meaning that these mutations do not spread to invasive areas. However, invasive populations have increased copy numbers of P450 genes compared with native populations through copy number variations. We also observed that invasive populations had increased proportions of resistance mutations at AChE. These results can be best explained by selective pressure causing increased resistance against synthetic insecticides in invasive populations.

Bioassay experiments demonstrated that Chinese populations are susceptible to Bt Cry1Fa toxins [31]. The analyzed Chinese samples in this study did not have mutations causing Cry1Fa resistance, explaining the susceptibility to this Bt toxin. Interestingly, a South African population showed moderate resistance against Bt Cry1A [32], while GC insertion causing Cry1A resistance [18,19] was not identified from invasive populations in the resequencing data. Therefore, the South African population could potentially have a resistance mutation that is yet to be characterized in native populations, and we cannot exclude the possibility of on-going gene flow of unidentified resistance mutations either from native to invasive populations or from other invasive populations that represent independent introduction from South African FAWs. More studies are necessary to identify causal resistance mutation in the South African populations and to identify the origin of this resistance mutation.

If a resistance Bt mutation exists in invasive populations with low frequency, individuals with the resistance mutations might not be included in the analyzed resequencing data. If this is true, existing resistance mutations in invasive populations might be undetected. The probability of not finding a resistance mutation is $(1 - p)^{2N}$, where p is the allele frequency of resistance mutations and N is the number of diploid organisms. If the allele frequency of resistance Bt mutation is 0.01, representing a rare allele, the probability of missing this mutation in our resequencing data is only $(1 - 0.01)^{2 \times 99} = 13.67\%$. Therefore, we do not believe that a resistance Bt mutation is likely to exist. Even in the case that this possibility of 13.67% is true, the resistance mutation is likely to be eliminated in a population by genetic drift due to the low allele frequency (0.01).

A whole-genome analysis is necessary to contrast the abundance of P450 genes between invasive and native populations, because P450 genes exist in multi-copies scattered across the genome [9–11]. Our whole genome analysis at a population level shows an increased copy number of P450 genes in the invasive FAW populations. Bioassay experiments demonstrated that Chinese populations have resistance to pyrethroid insecticides [32,45]. As P450 genes are known to cause resistance against deltamethrin [23], a type of pyrethroid in FAWs, the increased P450 gene numbers might be responsible for this resistance. Chinese populations also have resistance against organophosphate [32,45] as well as oxadiazine and diamide [32]. As P450 proteins are probably able to detoxify other types of insecticides than pyrethroid, the increased P450 genes may be responsible for the resistance against a wider range of insecticides.

In addition, the increased proportion of resistance mutations at AChE can cause the increased resistance against synthetic insecticides in invasive populations. Multiple resistance mutations with the increased proportion imply strong selective pressure for resistance against synthetic insecticide. The vast majority of beneficial mutations are quickly removed in a population due to stochastic fluctuation of allele frequency by the genetic bottleneck, and the survival probability of beneficial mutations is just two times that of the selective coefficient. Therefore, the probability of observing a single resistance mutation at AChE is proportional to the extent of increased fitness due to the resistance. The probability of observing multiple resistance mutations is a function of the multiplication of each selective coefficient, ranging between 0 to 1, at each resistance mutation. Therefore, multiple resistance mutations are difficult to observe without very high selective coefficients, and we argue that invasive populations are under strong selective pressure for the resistance against synthetic insecticides.

Glutathione S transferases or UDP glycosyltransferase might also cause resistance against synthetic insecticides. However, the direct relationship between these two gene families and resistance is yet to be tested in FAW. Genomic analyses and bioassay experiments will generate a more comprehensive list of causal resistance mutations against synthetic insecticides.

The differential usage of synthetic insecticides may explain the observations that invasive populations have increased allele frequency of resistance mutations at AchE genes and increased number of P450 genes by CNVs. The average annual insecticide use is particularly high in East Asia including China [46]. We speculate that initially introduced FAW insects in East Asia might have survived through adaptive evolution by the increased P450 gene number or the increased allele frequency of resistance mutations at AchE genes, while these FAWs remained undetected due to the low population sizes before explosive population expansion in West Africa, where genetic admixture occurred between populations with different invasive origins [31]. More studies are necessary to test the role of adaptive evolution associated with synthetic insecticide resistance in invasive success.

The F_{ST} result supports natural selection for Bt resistance in the Brazilian population. If selective pressure with the same directionality exists in invasive populations, newly generated Bt resistance mutations can be selected in the near future. Theoretically, species with large population sizes have higher adaptive potential than species with small population sizes because of a higher population-scaled rate of mutation [47,48]. Since the estimated population size of FAWs is as high as 3.75 million [49], the population-scaled mutation rate can be sufficient to generate a genetic variation for drastically rapid Darwinian positive selection. We observed two synonymous codons at a resistance position in AChE from a population Benin, implying that mutational influx could not be a limiting factor of adaptive evolution through positive selection. Therefore, invasive populations might experience adaptive evolution, causing resistance against Bt mutations in the near future.

5. Conclusions

In this study, we showed that Bt resistance mutations did not spread during this period to the Old World by the surveyed invasive FAW populations and that invasive populations have increased copy numbers of P450 genes and increased proportions of resistance mutations at AchE compared with native populations. These results imply that invasive populations might be under selective pressure for the resistance against synthetic insecticides.

The samples used in this study were obtained between 2017 to 2018. Since the first reported FAWs in Western Africa in 2016 [31], the genetic information in this study represents snapshots at the early phase of invasions. As Bt crops and synthetic insecticides are commonly used in the invasive area, the level of insecticide resistance can be changed together with underlying genotypes. This change is likely because our result is in line with selective pressure both for Bt and, in particular, synthetic insecticides. Therefore, monitoring changes in field susceptibility to various insecticides will be helpful in FAW control, together with the analysis of changes in genomics sequences to identify known or new causal resistance mutations.

Supplementary Materials: The following are available online at https://www.mdpi.com/article/10.3390/insects12050468/s1 (Table S1: Sample collection site and year, Table S2: The list of analyzed P450 genes in this study).

Author Contributions: Conceptualization, S.Y., T.W., and K.N.; Formal analysis, S.Y. and K.N.; Resources, N.N., P.J.S., T.B., W.T.T., K.G., an E.d.; Writing—original draft preparation, S.Y. and K.N.; Writing—review and editing, N.N., P.J.S., T.B., W.T.T., K.G., E.d., and T.W.; Supervision, K.N., T.B. and E.d.; Funding acquisition, K.N. All authors have read and agreed to the published version of the manuscript.

Funding: This work (ID 1702-018, given to K.N.) was publicly funded through ANR (the French National Research Agency) under the "Investissements d'avenir" program with the reference ANR-10-LABX-001-01 Labex Agro and coordinated by Agropolis Fondation under the frame of I-SITE MUSE (ANR-16-IDEX-0006). In addition, a grant from the department of Santé des Plantes et Environnement at Institut national de la recherche agronomique was awarded to K.N. (adaptivesv). This work was also financially supported by EUPHRESCO (FAW-spedcom, given to Anne-Nathalie Volkoff) and by CSIRO Health & Biosecurity (given to W.T.T., T.W., and K.G.). S.Y. was supported by a CIRAD-INRAE PhD fellowship.

Institutional Review Board Statement: Not applicable.

Informed Consent Statement: Not applicable.

Data Availability Statement: We will provide all files generated in this study upon request. These files include vcf files, gene annotation files, and computer programming scripts.

Conflicts of Interest: The authors declare no conflict of interest.

References

1. Tabashnik, B.E.; Brévault, T.; Carrière, Y. Insect Resistance to Bt Crops: Lessons from the First Billion Acres. *Nat. Biotechnol.* **2013**, *31*, 510–521. [CrossRef]
2. Jurat-Fuentes, J.L.; Heckel, D.G.; Ferré, J. Mechanisms of Resistance to Insecticidal Proteins from *Bacillus thuringiensis*. *Annu. Rev. Entomol.* **2021**, *66*, 121–140. [CrossRef]
3. Hemingway, J.; Field, L.; Vontas, J. An Overview of Insecticide Resistance. *Science* **2002**, *298*, 96–97. [CrossRef]
4. Carvalho, R.A.; Omoto, C.; Field, L.M.; Williamson, M.S.; Bass, C. Investigating the Molecular Mechanisms of Organophosphate and Pyrethroid Resistance in the Fall Armyworm *Spodoptera frugiperda*. *PLoS ONE* **2013**, *8*, e62268. [CrossRef]
5. Shan, T.; Chen, C.; Ding, Q.; Chen, X.; Zhang, H.; Chen, A.; Shi, X.; Gao, X. Molecular Characterization and Expression Profiles of Nicotinic Acetylcholine Receptors in *Bradysia odoriphaga*. *Pestic. Biochem. Physiol.* **2020**, *165*, 104563. [CrossRef] [PubMed]
6. Kim, A.-Y.; Kwon, D.H.; Jeong, I.H.; Koh, Y.H. An Investigation of the Molecular and Biochemical Basis Underlying Chlorantraniliprole-Resistant *Drosophila* Strains and Their Cross-Resistance to Other Insecticides. *Arch. Insect Biochem. Physiol.* **2018**, *99*, e21514. [CrossRef]
7. Lin, L.; Hao, Z.; Cao, P.; Yuchi, Z. Homology Modeling and Docking Study of Diamondback Moth Ryanodine Receptor Reveals the Mechanisms for Channel Activation, Insecticide Binding and Resistance. *Pest Manag. Sci.* **2020**, *76*, 1291–1303. [CrossRef]
8. Montezano, D.G.; Specht, A.; Sosa-Gómez, D.R.; Roque-Specht, V.F.; Sousa-Silva, J.C.; de Paula-Moraes, S.V.; Peterson, J.A.; Hunt, T.E. Host Plants of *Spodoptera frugiperda* (Lepidoptera: Noctuidae) in the Americas. *Afr. Entomol.* **2018**, *26*, 286–300. [CrossRef]
9. Gouin, A.; Bretaudeau, A.; Nam, K.; Gimenez, S.; Aury, J.-M.; Duvic, B.; Hilliou, F.; Durand, N.; Montagné, N.; Darboux, I.; et al. Two Genomes of Highly Polyphagous Lepidopteran Pests (*Spodoptera frugiperda*, Noctuidae) with Different Host-Plant Ranges. *Sci. Rep.* **2017**, *7*, 11816. [CrossRef]
10. Gui, F.; Lan, T.; Zhao, Y.; Guo, W.; Dong, Y.; Fang, D.; Liu, H.; Li, H.; Wang, H.; Hao, R. Genomic and Transcriptomic Analysis Unveils Population Evolution and Development of Pesticide Resistance in Fall Armyworm *Spodoptera frugiperda*. *Protein Cell* **2020**. [CrossRef]
11. Xiao, H.; Ye, X.; Xu, H.; Mei, Y.; Yang, Y.; Chen, X.; Yang, Y.; Liu, T.; Yu, Y.; Yang, W. The Genetic Adaptations of Fall Armyworm *Spodoptera frugiperda* Facilitated Its Rapid Global Dispersal and Invasion. *Mol. Ecol. Resour.* **2020**, *20*, 1050–1068. [CrossRef] [PubMed]
12. Jakka, S.R.K.; Gong, L.; Hasler, J.; Banerjee, R.; Sheets, J.J.; Narva, K.; Blanco, C.A.; Jurat-Fuentes, J.L. Field-Evolved Mode 1 Resistance of the Fall Armyworm to Transgenic Cry1Fa-Expressing Corn Associated with Reduced Cry1Fa Toxin Binding and Midgut Alkaline Phosphatase Expression. *Appl. Environ. Microbiol.* **2016**, *82*, 1023–1034. [CrossRef] [PubMed]
13. Storer, N.P.; Babcock, J.M.; Schlenz, M.; Meade, T.; Thompson, G.D.; Bing, J.W.; Huckaba, R.M. Discovery and Characterization of Field Resistance to Bt Maize: *Spodoptera frugiperda* (Lepidoptera: Noctuidae) in Puerto Rico. *J. Econ. Entomol.* **2010**, *103*, 1031–1038. [CrossRef]
14. Monnerat, R.; Martins, E.; Macedo, C.; Queiroz, P.; Praça, L.; Soares, C.M.; Moreira, H.; Grisi, I.; Silva, J.; Soberon, M.; et al. Evidence of Field-Evolved Resistance of *Spodoptera frugiperda* to Bt Corn Expressing Cry1F in Brazil That Is Still Sensitive to Modified Bt Toxins. *PLoS ONE* **2015**, *10*, e0119544. [CrossRef]

15. Omoto, C.; Bernardi, O.; Salmeron, E.; Sorgatto, R.J.; Dourado, P.M.; Crivellari, A.; Carvalho, R.A.; Willse, A.; Martinelli, S.; Head, G.P. Field-Evolved Resistance to Cry1Ab Maize by *Spodoptera frugiperda* in Brazil. *Pest Manag. Sci.* **2016**, *72*, 1727–1736. [CrossRef] [PubMed]

16. Chandrasena, D.I.; Signorini, A.M.; Abratti, G.; Storer, N.P.; Olaciregui, M.L.; Alves, A.P.; Pilcher, C.D. Characterization of Field-Evolved Resistance to *Bacillus Thuringiensis*-Derived Cry1F δ-Endotoxin in *Spodoptera frugiperda* Populations from Argentina. *Pest Manag. Sci.* **2018**, *74*, 746–754. [CrossRef]

17. Schlum, K.A.; Lamour, K.; de Bortoli, C.P.; Banerjee, R.; Meagher, R.; Pereira, E.; Murua, M.G.; Sword, G.A.; Tessnow, A.E.; Viteri Dillon, D.; et al. Whole Genome Comparisons Reveal Panmixia among Fall Armyworm (*Spodoptera frugiperda*) from Diverse Locations. *BMC Genom.* **2021**, *22*, 179. [CrossRef]

18. Banerjee, R.; Hasler, J.; Meagher, R.; Nagoshi, R.; Hietala, L.; Huang, F.; Narva, K.; Jurat-Fuentes, J.L. Mechanism and DNA-Based Detection of Field-Evolved Resistance to Transgenic Bt Corn in Fall Armyworm (*Spodoptera frugiperda*). *Sci. Rep.* **2017**, *7*, 10877. [CrossRef]

19. Flagel, L.; Lee, Y.W.; Wanjugi, H.; Swarup, S.; Brown, A.; Wang, J.; Kraft, E.; Greenplate, J.; Simmons, J.; Adams, N.; et al. Mutational Disruption of the ABCC2 Gene in Fall Armyworm, *Spodoptera frugiperda*, Confers Resistance to the Cry1Fa and Cry1A.105 Insecticidal Proteins. *Sci. Rep.* **2018**, *8*, 7255. [CrossRef]

20. Boaventura, D.; Ulrich, J.; Lueke, B.; Bolzan, A.; Okuma, D.; Gutbrod, O.; Geibel, S.; Zeng, Q.; Dourado, P.M.; Martinelli, S.; et al. Molecular Characterization of Cry1F Resistance in Fall Armyworm, *Spodoptera frugiperda* from Brazil. *Insect Biochem. Mol. Biol.* **2020**, *116*, 103280. [CrossRef]

21. Guan, F.; Zhang, J.; Shen, H.; Wang, X.; Padovan, A.; Walsh, T.K.; Tay, W.T.; Gordon, K.H.J.; James, W.; Czepak, C.; et al. Whole-Genome Sequencing to Detect Mutations Associated with Resistance to Insecticides and Bt Proteins in *Spodoptera frugiperda*. *Insect Sci.* **2020**. [CrossRef] [PubMed]

22. Gutiérrez-Moreno, R.; Mota-Sanchez, D.; Blanco, C.A.; Whalon, M.E.; Terán-Santofimio, H.; Rodriguez-Maciel, J.C.; DiFonzo, C. Field-Evolved Resistance of the Fall Armyworm (Lepidoptera: Noctuidae) to Synthetic Insecticides in Puerto Rico and Mexico. *J. Econ. Entomol.* **2019**, *112*, 792–802. [CrossRef] [PubMed]

23. Gimenez, S.; Abdelgaffar, H.; Goff, G.L.; Hilliou, F.; Blanco, C.A.; Hänniger, S.; Bretaudeau, A.; Legeai, F.; Nègre, N.; Jurat-Fuentes, J.L.; et al. Adaptation by Copy Number Variation Increases Insecticide Resistance in the Fall Armyworm. *Commun. Biol.* **2020**, *3*, 664. [CrossRef] [PubMed]

24. Boaventura, D.; Bolzan, A.; Padovez, F.E.; Okuma, D.M.; Omoto, C.; Nauen, R. Detection of a Ryanodine Receptor Target-Site Mutation in Diamide Insecticide Resistant Fall Armyworm, *Spodoptera frugiperda*. *Pest Manag. Sci.* **2020**, *76*, 47–54. [CrossRef]

25. Goergen, G.; Kumar, P.L.; Sankung, S.B.; Togola, A.; Tamò, M. First Report of Outbreaks of the Fall Armyworm *Spodoptera frugiperda* (J E Smith) (Lepidoptera, Noctuidae), a New Alien Invasive Pest in West and Central Africa. *PLoS ONE* **2016**, *11*, e0165632. [CrossRef]

26. Day, R.; Abrahams, P.; Bateman, M.; Beale, T.; Clottey, V.; Cock, M.; Colmenarez, Y.; Corniani, N.; Early, R.; Godwin, J. Fall Armyworm: Impacts and Implications for Africa. *Outlooks Pest Manag.* **2017**, *28*, 196–201. [CrossRef]

27. Pashley, D.P. Host-Associated Genetic Differentiation in Fall Armyworm (Lepidoptera: Noctuidae): A Sibling Species Complex? *Ann. Entomol. Soc. Am.* **1986**, *79*, 898–904. [CrossRef]

28. Pashley, D.P. Host-associated differentiation in armyworms (Lepidoptera: Noctuidae): An allozymic and mitochondrial DNA perspective. In *Electrophoretic Studies on Agricultural Pests*; Loxdale, H.D., Hollander, J.D., Eds.; Clarendon Press: Oxford, UK, 1989; Volume 39, pp. 103–144.

29. Dumas, P.; Legeai, F.; Lemaitre, C.; Scaon, E.; Orsucci, M.; Labadie, K.; Gimenez, S.; Clamens, A.-L.; Henri, H.; Vavre, F.; et al. *Spodoptera frugiperda* (Lepidoptera: Noctuidae) Host-Plant Variants: Two Host Strains or Two Distinct Species? *Genetica* **2015**, *143*, 305–316. [CrossRef]

30. Nagoshi, R.N.; Goergen, G.; Plessis, H.D.; van den Berg, J.; Meagher, R. Genetic Comparisons of Fall Armyworm Populations from 11 Countries Spanning Sub-Saharan Africa Provide Insights into Strain Composition and Migratory Behaviors. *Sci. Rep.* **2019**, *9*, 8311. [CrossRef]

31. Yainna, S.; Tay, W.T.; Fiteni, E.; Legeai, F.; Clamens, A.-L.; Gimenez, S.; Frayssinet, M.; Asokan, R.; Kalleshwaraswamy, C.M.; Deshmukh, S.; et al. Genomic Balancing Selection Is Key to the Invasive Success of the Fall Armyworm. *bioRxiv* **2020**. [CrossRef]

32. Zhang, L.; Liu, B.; Zheng, W.; Liu, C.; Zhang, D.; Zhao, S.; Li, Z.; Xu, P.; Wilson, K.; Withers, A.; et al. Genetic Structure and Insecticide Resistance Characteristics of Fall Armyworm Populations Invading China. *Mol. Ecol. Resour.* **2020**, *20*, 1682–1696. [CrossRef] [PubMed]

33. Tay, W.T.; Rane, R.; Padovan, A.; Walsh, T.; Elfekih, S.; Downes, S.; Nam, K.; d'Alençon, E.; Zhang, J.; Wu, Y.; et al. Whole Genome Sequencing of Global *Spodoptera frugiperda* Populations: Evidence for Complex, Multiple Introductions across the Old World. *bioRxiv* **2020**. [CrossRef]

34. Boaventura, D.; Martin, M.; Pozzebon, A.; Mota-Sanchez, D.; Nauen, R. Monitoring of Target-Site Mutations Conferring Insecticide Resistance in *Spodoptera frugiperda*. *Insects* **2020**, *11*, 545. [CrossRef]

35. McKenna, A.; Hanna, M.; Banks, E.; Sivachenko, A.; Cibulskis, K.; Kernytsky, A.; Garimella, K.; Altshuler, D.; Gabriel, S.; Daly, M.; et al. The Genome Analysis Toolkit: A MapReduce Framework for Analyzing next-Generation DNA Sequencing Data. *Genome Res.* **2010**, *20*, 1297–1303. [CrossRef] [PubMed]

36. Wang, X.; Zheng, Z.; Cai, Y.; Chen, T.; Li, C.; Fu, W.; Jiang, Y. CNVcaller: Highly Efficient and Widely Applicable Software for Detecting Copy Number Variations in Large Populations. *GigaScience* **2017**, *6*. [CrossRef]

37. Create Your Own Custom Map. Available online: https://mapchart.net/index.html (accessed on 17 May 2021).

38. Slater, G.S.C.; Birney, E. Automated Generation of Heuristics for Biological Sequence Comparison. *BMC Bioinform.* **2005**, *6*, 3 [CrossRef] [PubMed]

39. Robinson, J.T.; Thorvaldsdóttir, H.; Winckler, W.; Guttman, M.; Lander, E.S.; Getz, G.; Mesirov, J.P. Integrative Genomics Viewer. *Nat. Biotechnol.* **2011**, *29*, 24–26. [CrossRef]

40. Camacho, C.; Coulouris, G.; Avagyan, V.; Ma, N.; Papadopoulos, J.; Bealer, K.; Madden, T.L. BLAST+: Architecture and Applications. *BMC Bioinform.* **2009**, *10*, 421. [CrossRef]

41. Weir, B.S.; Cockerham, C.C. Estimating F-Statistics for the Analysis of Population Structure. *Evolution* **1984**, *38*, 1358–1370.

42. Danecek, P.; Auton, A.; Abecasis, G.; Albers, C.A.; Banks, E.; DePristo, M.A.; Handsaker, R.E.; Lunter, G.; Marth, G.T.; Sherry, S.T.; et al. The Variant Call Format and VCFtools. *Bioinformatics* **2011**, *27*, 2156–2158. [CrossRef]

43. Walsh, T.K.; Joussen, N.; Tian, K.; McGaughran, A.; Anderson, C.J.; Qiu, X.; Ahn, S.-J.; Bird, L.; Pavlidi, N.; Vontas, J.; et al. Multiple Recombination Events between Two Cytochrome P450 Loci Contribute to Global Pyrethroid Resistance in *Helicoverpa armigera*. *PLoS ONE* **2018**, *13*, e0197760. [CrossRef] [PubMed]

44. Valencia-Montoya, W.A.; Elfekih, S.; North, H.L.; Meier, J.I.; Warren, I.A.; Tay, W.T.; Gordon, K.H.; Specht, A.; Paula-Moraes, S.V.; Rane, R. Adaptive Introgression across Semipermeable Species Boundaries between Local *Helicoverpa zea* and Invasive *Helicoverpa armigera* Moths. *Mol. Biol. Evol.* **2020**, *37*, 2568–2583. [CrossRef] [PubMed]

45. Zhang, D.; Xiao, Y.; Xu, P.; Yang, X.; Wu, Q.; Wu, K. Insecticide Resistance Monitoring for the Invasive Populations of Fall Armyworm, *Spodoptera frugiperda* in China. *J. Integr. Agric.* **2021**, *20*, 783–791. [CrossRef]

46. Zhang, W. Global Pesticide Use: Profile, Trend, Cost/Benefit and More. *Proc. Int. Acad. Ecol. Environ. Sci.* **2018**, *8*, 1–27.

47. Lanfear, R.; Kokko, H.; Eyre-Walker, A. Population Size and the Rate of Evolution. *Trends Ecol. Evol.* **2014**, *29*, 33–41. [CrossRef] [PubMed]

48. Nam, K.; Munch, K.; Mailund, T.; Nater, A.; Greminger, M.P.; Krützen, M.; Marquès-Bonet, T.; Schierup, M.H. Evidence That the Rate of Strong Selective Sweeps Increases with Population Size in the Great Apes. *Proc. Natl. Acad. Sci. USA* **2017**, *114*, 1613–1618. [CrossRef] [PubMed]

49. Nam, K.; Nhim, S.; Robin, S.; Bretaudeau, A.; Nègre, N.; d'Alençon, E. Positive Selection Alone Is Sufficient for Whole Genome Differentiation at the Early Stage of Speciation Process in the Fall Armyworm. *BMC Evol. Biol.* **2020**, *20*, 152. [CrossRef]

 *insects*

MDPI

Communication

Evaluation of Reference Genes and Expression Level of Genes Potentially Involved in the Mode of Action of Cry1Ac and Cry1F in a Susceptible Reference Strain of *Chrysodeixis includens*

Macarena Martin [1], Debora Boaventura [2] and Ralf Nauen [2,*]

1   Department of Agronomy, Food, Natural Resources, Animals and Environment, University of Padova, 35020 Padova, Italy; macamar_20@hotmail.com
2   Bayer AG, Crop Science Division, R&D Pest Control, Alfred Nobel Str. 50, 40789 Monheim, Germany; debora.duarteboaventura@bayer.com
*   Correspondence: ralf.nauen@bayer.com

Simple Summary: Soybean looper (a moth species) is a major pest of soybean plants in the American continent and its larvae need to be kept under economic damage thresholds to guarantee sustainable yields. Soybean looper control relies mostly on the use of insecticides and genetically modified crops expressing *Bacillus thuringiensis* (Bt) insecticidal proteins. Due to the high selection pressure exerted by these control measures, resistance has developed to different insecticides and Bt proteins. Here, we tested the basal sensitivity of a soybean looper laboratory reference strain against two insecticidal proteins and determined the level of expression of potential receptors of these proteins across all (six) larval stages. Furthermore, we identified stable reference genes across all larval stages to normalize gene expression data obtained by quantitative polymerase chain reaction (qPCR). The results presented in this communication are useful to support future studies on insecticide and insecticidal protein resistance in soybean looper.

Abstract: Soybean looper (SBL), *Chrysodeixis includens* (Walker), is one of the major lepidopteran pests of soybean in the American continent. SBL control relies mostly on the use of insecticides and genetically modified crops expressing *Bacillus thuringiensis* (Bt) insecticidal Cry proteins. Due to the high selection pressure exerted by these control measures, resistance has developed to different insecticides and Bt proteins. Nevertheless, studies on the mechanistic background are still scarce. Here, the susceptibility of the laboratory SBL-Benzon strain to the Bt proteins Cry1Ac and Cry1F was determined in diet overlay assays and revealed a greater activity of Cry1Ac than Cry1F, thus confirming results obtained for other sensitive SBL strains. A reference gene study across larval stages with four candidate genes revealed that *RPL10* and *EF1* were the most stable genes for normalization of gene expression data obtained by RT-qPCR. Finally, the basal expression levels of eight potential Bt protein receptor genes in six larval instars were analyzed, including ATP-binding cassette (*ABC*) transporters, alkaline phosphatase, aminopeptidases, and cadherin. The results presented here provide fundamental knowledge to support future SBL resistance studies.

Keywords: soybean looper; *Bacillus thuringiensis*; resistance; reference genes; *ABCC2*

 check for
updates

Citation: Martin, M.; Boaventura, D.; Nauen, R. Evaluation of Reference Genes and Expression Level of Genes Potentially Involved in the Mode of Action of Cry1Ac and Cry1F in a Susceptible Reference Strain of *Chrysodeixis includens*. *Insects* **2021**, *12*, 598. https://doi.org/10.3390/insects12070598

Academic Editor: Yulin Gao

Received: 2 June 2021
Accepted: 29 June 2021
Published: 30 June 2021

Publisher's Note: MDPI stays neutral with regard to jurisdictional claims in published maps and institutional affiliations.

1. Introduction

Soybean looper (SBL), *Chrysodeixis includens* (Walker) (Lepidoptera: Noctuidae), is a distinctive species from the Western hemisphere and a key pest of soybean in many countries of the American continent [1]. The control of SBL relies on genetically modified crops expressing *Bacillus thuringiensis* (Bt) insecticidal proteins and foliar insecticide applications [2].

Frequent insecticide applications have led to insecticide resistance against some major chemical classes, such as pyrethroids, carbamates, organophosphates, benzoylureas, and diamides [3–5]. Moreover, the high selection pressure exerted by millions of hectares of soybean expressing Bt proteins has resulted in resistance towards Cry1Ac [6].

A reduction in Cry protein binding due to changes in the expression and/or mutations in midgut receptors involved in Cry protein-mediated intoxication was the most frequent mechanism of resistance [7]. Different proteins have been identified as receptors for Cry proteins in lepidopteran species, including aminopeptidases (*APN*), membrane bound alkaline phosphatases (m*ALP*), cadherins (*CAD*), and ATP-binding cassette (*ABC*) transporters [8].

Brazilian field populations of SBL selected against Cry1Ac in the laboratory showed up to 127-fold resistance [6]. However, the molecular mechanism involved in the resistance of SBL towards Cry1Ac is still unknown. In other lepidopteran pest species, resistance to Cry1Ac has been associated with mutations and/or differences in the expression levels of *ABCC2*, *APN*, m*ALP*, and *CAD* [9–13].

Here, we have tested the basal sensitivity of the commonly known reference SBL Benzon strain towards two major Bt proteins, Cry1Ac and Cry1F. Moreover, we have investigated, for the first time, the stability of potential reference genes with low expression variance across larval instars of SBL for normalization of RT-qPCR expression data. As resistance towards Bt proteins in other species has been linked to differences in the expression of potential Bt receptor genes, we have analyzed the basal expression of eight potential Bt protein receptor genes in six larval instars to support the characterization of resistance mechanisms to Cry toxins in SBL in future studies.

2. Material and Methods

2.1. Insects

SBL-Benzon strain from Benzon Research Inc., USA was reared in the laboratory under controlled conditions (25 ± 1°C, 55 ± 5% relative humidity, and a photoperiod of L16:D8). Briefly, adults were kept in cages covered with filter paper for egg laying and fed with honey water pads every second day. Larvae were mass-kept in a plastic box (18 × 13 × 5 cm) containing Soybean-Looper-Diet (Southland Products Inc., Lake Village, AR, USA) until the third instar. Afterwards, two third-instar larvae were confined to petri dishes (diameter 6 cm) containing the same diet until pupation.

2.2. Cry1Ac and Cry1F bioassays

The bioassays were performed according to Boaventura et al. 2020 [14]. Briefly, seven concentrations of Cry1Ac (450-0.62 ng cm^{-2}) and eight concentrations of Cry1F (12,150-5.56 ng cm^{-2}) were tested. Dilutions were prepared from Cry1Ac lyophilized material (28.2% (*w/w*) purity) and Cry1F (TIC-842) crystal spore solution [15] in 50 mM sodium carbonate buffer (pH 10.4) and 0.1% (*v/v*) Triton X-100. Twenty-five µL of Bt solution was applied to the surface of artificial diet in a 48-well plate (Greiner CELLSTAR®, Merck, Darmstadt, Germany) and a single SBL neonate larva (<24 h old) was placed into each well and kept at 20 ± 2 °C.

A total number of 12 neonates were used per protein concentration and control (buffer and 0.1% (*v/v*) Triton X-100). For the highest concentration, only six neonates were used and the other six neonates were placed in wells containing only artificial diet. The bioassay was replicated three times and the mortality was scored after five days; larvae that did not move or reach the third instar were considered dead. The results were fitted by a logistic regression model to calculate LC$_{50}$/LC$_{95}$ values and 95% confidence intervals (GraphPad Software Inc. v.5, San Diego, CA, USA).

2.3. RNA Extraction and cDNA Synthesis

Total RNA from each instar was extracted from 30 pooled (first instar) or 10 pooled larvae (second to sixth instar) of SBL in four biological replicates. The RNA extraction from first and second instars was performed with the PicoPure™ RNA Isolation Kit (Thermo Fisher, Vilnius, Lithuania) according to the manufacturer's instructions. For the third to sixth instar, the larval tissue was ground with a Retsch TissueLyser (QIAGEN, Hilden, Germany) or mortar and pestle. The total RNA was extracted from powdered samples using TRIzol® reagent (Invitrogen, Waltham, MA, USA) followed by RNA purification according to RNeasy® Plus Universal Mini Kit (QIAGEN, Hilden, Germany) recommendations including gDNA elimination with DNase I. The RNA concentration was normalized to 100 ng μL^{-1} and 1 µg total RNA was used in 20 µL reactions for cDNA synthesis using iScript™ cDNA synthesis (Bio-Rad, Laboratories, CA, USA) according to the manufacturer's instructions.

2.4. Screening for Reference Genes

The expression of four candidate reference genes was tested: *ribosomal protein L10 (RPL10)*, *glyceraldehyde-3-phosphate dehydrogenase (GAPDH)*, *elongation factor-1 alpha (EF1)*, and *beta actin 1 (ACT1)*. The reference gene sequences were obtained from a custom-made SBL transcriptome generated at Bayer AG through queries using local BLAST searches and submitted to GenBank (Table S1). The primers were designed using Geneious software v. 10.2.3 (Biomatters Ltd, Auckland, New Zealand) (Table S1). The RT-qPCR reactions were performed as recently described [14] using reverse/forward primers shown in Table S1. Amplification efficiencies were determined by a series of 5-fold dilutions to create standard curves by a linear regression model [16] (Table S1).

The cycle threshold values (Ct values) were obtained from CFX Maestro 1.0 v4.0 (Bio-Rad) software. The expression stability of four candidate reference genes was analyzed using geNorm qBase Plus v3.1 software (Biogazelle, Belgium) and NormFinder (moma.dk/normfinder-software) using stability value (M) below 0.5 and pairwise variation (V) value below 0.15 [16]. The Excel-based tool Normfinder was used according to Andersen et al. (2004) [17].

2.5. Expression Analysis of Possible Bt Protein Target Genes

The expression level of eight potential Bt receptor genes (*ABCC2*, *ABCC3*, *CAD*, *mALP*, and *APN* (*APN1-4*)) was determined by RT-qPCR in six larval instars. The genes *EF1* and *RPL10* were the most stable genes across SBL larval instars and were used for normalization. In total, four biological replicates containing cDNA from a pool of 10-30 larvae, as described in Section 2.3, were used for analyzing the normalized expression levels. Primer pairs are described in Table S1 and the RT-qPCR conditions were chosen as recently described [14]. The normalized expression values obtained for the different developmental stages were compared to the Ct values from the first instar and statistically tested by one-way ANOVA, followed by Tukey–Kramer multiple-comparison test with qbase + software (v. 3.2; Biogazelle, Belgium).

3. Results

3.1. Bioassays

For Cry1Ac, the LC_{50} and LC_{95} values were 10.41 ng cm^{-2} and 44.56 ng cm^{-2}, respectively. Cry1F showed an LC_{50} of 228.1 ng cm^{-2} and an LC_{95} of 4,967 ng cm^{-2} (Figure 1, Table S2).

3.2. Screening for Reference Genes

According to geNorm, *RPL10* was identified as the most stable gene across larval stages of SBL (M = 0.458), followed by *EF1*, *GAPDH*, and *ACT1* with M values of 0.519, 0.559, and 0.637, respectively (Table 1). When using NormFinder, *EF1* (stability value = 0.242) and *RPL10* (0.244) were the most stable genes, followed by *GAPDH* (0.309) and *ACT1* (0.351) (Table 1).

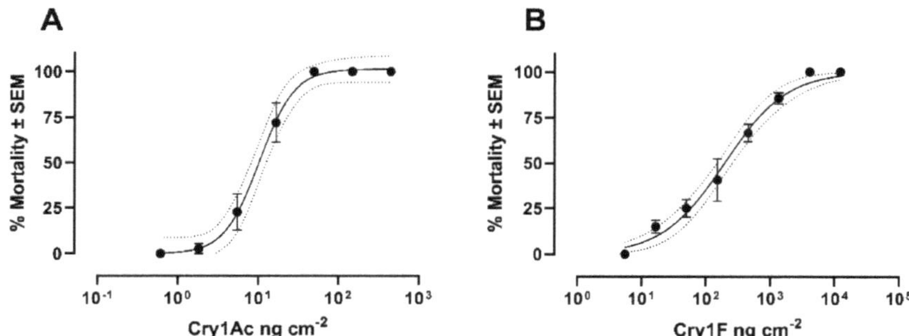

Figure 1. Toxicity of Bt proteins against neonates of soybean looper strain SBL-Benzon. **(A)** Toxicity of Cry1Ac (N = 288) and **(B)** toxicity of Cry1F (N = 324). Dashed lines indicate the 95% confidence band.

Table 1. Stability analysis of the candidate reference genes *ribosomal protein L10 (RPL10), glyceraldehyde 3-phosphate dehydrogenase (GAPDH), elongation factor-1 alpha (EF1),* and *beta actin 1 (ACT1)* analyzed by geNorm (average expression stability or M value) and NormFinder (stability value) in different larval stages of *Chrysodeixis includens.*

Method	EF1	RPL10	GAPDH	ACT1
geNorm	0.519	0.458	0.559	0.637
NormFinder	0.242	0.244	0.309	0.351

3.3. Expression Analysis of Potential Bt Protein Target Genes

The basal expression levels of the genes *ABCC2, ABCC3, CAD, mALP,* and *APN (APN1, 4)*, reported to be potential receptors of Cry1Ac and Cry1F in other lepidopteran species, were analyzed in the different larval stages of the Bt-susceptible SBL-Benzon strain. For *ABCC3*, no significant difference in the expression level across larval stages was observed with average normalized values between 1 and 2.02. In contrast, the expression level of *ABCC2* was significantly lower in second-instar larvae (p-value < 0.05) (Figure 2).

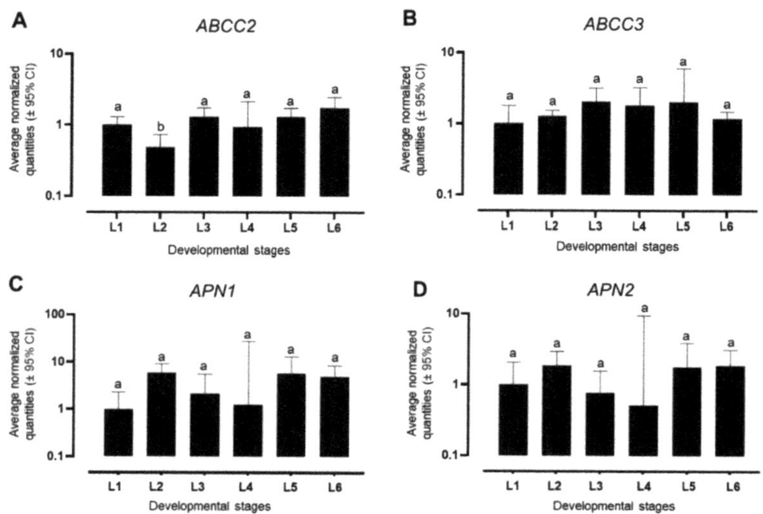

Figure 2. *Cont.*

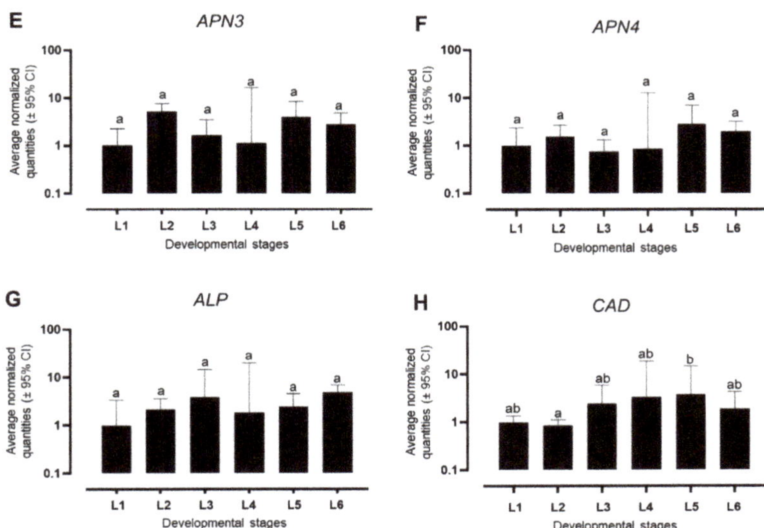

Figure 2. RT-qPCR expression analysis of genes possibly involved in Bt toxicity, analyzed by geNorm across different larval stages of *Chrysodeixis includens*. Average normalized quantities of (**A**) *ABCC2*, (**B**) *ABCC3*, (**C**) *APN1*, (**D**) *APN2*, (**E**) *APN3*, (**F**) *APN4*, (**G**) *ALP*, and (**H**) *CAD*. *CAD*: cadherin; *ABCC2/ABCC3*: ATP-binding cassette subfamily C2/C3 transporter; *ALP*: alkaline phosphatase; *APN*: amino peptidase class 1, 2, 3, and 4. Different letters denote a significant difference (One-way ANOVA; post hoc Tukey comparison, $p < 0.05$).

The expression level of the individual *APN* genes was not significantly different between the different larval stages. Furthermore, no significant differences in the expression level of m*ALP* were detected. However, the expression level of *CAD* in second-instar larvae was significantly lower than in fifth-instar larvae (Figure 2).

4. Discussion

The use of insecticide-susceptible reference strains such as SBL-Benzon is a key aspect in resistance monitoring and mechanism studies. SBL is a notorious pest of soybean crops, but it also feeds on cotton [18]. Cry1Ac is utilized in both crops and a relatively high survival rate of SBL in Cry1Ac Bt cotton has been reported [19]. The primary Insecticide Resistance Management (IRM) strategy for Bt crops has been the use of refuges (presence of non-Bt host plants near Bt crops). The success of refuge-based IRM depends on a low frequency of potential resistance alleles, high susceptibility to the expressed Bt protein(s), and a recessive mode of inheritance [20]. Utilizing Bt crops expressing more than one Bt protein (pyramid strategy), such as Bt soybean expressing Cry1Ac and Cry1F, is another valuable IRM strategy [21]. The combination of Cry1F and Cry1Ac in transgenic cotton was shown to be highly effective against SBL [22]. However, soybean and cotton are usually planted in succession in Brazil, thus increasing the resistance risk of SBL by continuous selection pressure, particularly against Cry1F, which exhibited significantly lower activity than Cry1Ac against the susceptible reference strain SBL-Benzon. The data obtained here were close to dose–response data published recently and showing the superior efficacy of Cry1Ac compared to Cry1F against neonates of a susceptible Brazilian SBL field strain [6]. This is in contrast to toxicity data previously published in another study, where Cry1F was more active than Cry1Ac against neonates of a sensitive SBL laboratory colony, whereas brush-border membrane vesicle (BBMV) binding studies revealed the opposite, i.e., a higher affinity of Cry1Ac than Cry1F [23].

The mechanisms driving Bt resistance are quite diverse and often associated with the differential expression of genes of potential Bt receptor proteins, particularly in noctuid pests reviewed in [24]. Bt protein resistance in SBL is not yet widespread, and mechanistic studies of factors driving resistance in SBL are scarce. However, the elucidation of mechanisms conferring Bt resistance based on differential gene expression analysis requires stable reference genes for normalization of RT-qPCR data [25]. Such studies have not yet been conducted in SBL, so we investigated the stability of four candidate reference genes across six larval instars to allow the selection of at least two appropriate reference genes typically considered sufficient for the normalization of RT-qPCR data to reduce any systematic bias [26]. Based on the results obtained in our study, *RPL10* and *EF1* were most stably expressed across larval stages of SBL and deemed suitable to be used in future gene expression studies. However, *EF1* was recently already employed for the normalization of the expression level of detoxification genes in a pyrethroid-resistant SBL strain [27].

Finally, we utilized the identified reference genes to normalize RT-qPCR data obtained for the basal expression level of potential Bt receptor genes in strain SBL-Benzon across six larval instars. Such data are a useful resource for the comparative analysis of gene expression levels in future studies with SBL strains resistant to Bt proteins. *ABCC2* has been described as an important receptor for Cry1Ac [10] and Cry1F [28] in lepidopteran pests, and mutations in *ABCC2* were shown to confer resistance to Cry1 proteins in *H. armigera* [10] and other noctuid pests such as fall armyworm [14,15,24]. Meanwhile, in *Plutella xylostella*, Cry1Ac resistance was linked to the downregulation of *ABCC2*, *ABCC3* and *mALP* [13]. However, in a *Helicoverpa zea* strain selected with Cry1Ac, resistance was associated with an increase in the expression of *ALP* [29]. Target-site mutations in *APN1* and differences in its expression level have been reported to cause resistance to Cry1Ac in *H. armigera* and *Trichoplusia ni,* respectively [9,11]. Cadherins were also shown to be receptors of Cry1Ac in different lepidopteran species [24,30]. Differences in cadherin levels and/or mutations in this gene were, for example, associated with Cry1Ac resistance in *Ostrinia furnacalis* [12]. Our results reveled no differences in *APN* and *ALP* expression levels across larval instars of Bt-susceptible SBL, but significant differences in *CAD* expression levels between the second and fifth larval instar; however, additional work will be necessary to investigate if such differences have implications for Bt protein toxicity.

In summary, we have shown the stability of the genes *RPL10* and *EF1* across six different larval stages of SBL and recommend their use for the normalization of gene expression data in future studies. In addition, we identified and described potential Bt target genes in SBL and investigated their basal expression profiles across larval instars in the reference SBL-Benzon strain to support future resistance studies in SBL.

5. Conclusions

Resistance towards Bt proteins has already been reported in SBL and the results presented in this study will help to support future research on SBL resistance towards both Bt proteins and synthetic insecticides. Besides the data presented on the susceptibility towards Cry1Ac and Cry1F proteins, this study provides useful insights into the stability of reference genes, with *RPL10* and *EF1* as the most stable ones. Moreover, for the first time, potential Bt target genes (*ABCC2*, *ABCC3*, *cadherin*, *mALP*, and several *APNs*) and their respective expression profiles were investigated for a susceptible SBL reference strain SBL-Benzon.

Supplementary Materials: The following are available online at https://www.mdpi.com/article/10.3390/insects12070598/s1, Table S1: Gene names of orthologous candidate reference genes (*RPL10, GADPH, EF1,* and *ACT1*), potential orthologous Cry toxin target genes, species, GenBank accession numbers and primer sequences for the amplification of the respective Chrysodeixis includens genes, Table S2: Toxicity (LC50-and LC95-values) of Cry1Ac and Cry1F to neonates of soybean looper Benzon strain and the respective goodness of fit calculated for the non-linear regressions used.

Author Contributions: R.N. and D.B. conceived the study. R.N. supervised the project. D.B. and M.M. performed the research and analyzed data. D.B., M.M. and R.N. wrote the paper. All authors have read and agreed to the published version of the manuscript.

Funding: M.M. thanks the Erasmus+–KA1 Erasmus Mundus Joint Master's Degree Programme of the European Commission as this paper has been developed as a result of a mobility stay funded by the PLANT HEALTH Project.

Institutional Review Board Statement: Not applicable.

Informed Consent Statement: Not applicable.

Data Availability Statement: The data presented in this study are available in the supplementary materials or on request from the corresponding author.

Conflicts of Interest: R.N. and D.B. are employed by Bayer AG, a manufacturer of insecticides and transgenic soybean. This work was partially financed by Bayer AG. The authors declare no additional conflict of interest.

References

1. Herzog, D.C. Sampling Soybean Looper on Soybean. In *Sampling Methods in Soybean Entomology*; Kogan, M., Herzog, D.C., Eds.; Springer: New York, NY, USA, 1980; pp. 141–168. [CrossRef]
2. Di Oliveira, J.R.G.; Ferreira, M.D.C.; Román, R.A.A. Diferentes diâmetros de gotas e equipamentos para aplicação de inseticida no controle de Pseudoplusia includens. *Eng. Agrícola* **2010**, *30*, 92–99. [CrossRef]
3. Boernel, D.J.; Mink, J.S.; Wier, A.T.; Thomas, J.D.; Leonard, B.R.; Gallardo, F. Management of Insecticide Resistant Soybean Loopers (Pseudoplusia Includens) in the Southern United States. In *Pest Management in Soybean*; Copping, L.G., Green, M.B.M., Rees, R.T., Eds.; Springer: Dordrecht, The Netherlands, 1992; pp. 66–87.
4. Owen, L.N.; Catchot, A.; Musser, F.; Gore, J.; Cook, D.; Jackson, R. Susceptibility ofChrysodeixis includens(Lepidoptera: Noctuidae) to Reduced-Risk Insecticides. *Fla. Èntomol.* **2013**, *96*, 554–559. [CrossRef]
5. Stacke, R.F.; Godoy, D.; Pretto, V.E.; Führ, F.M.; Gubiani, P.D.S.; Hettwer, B.L.; Garlet, C.G.; Somavilla, J.C.; Muraro, D.S.; Bernardi, O. Field-evolved resistance to chitin synthesis inhibitor insecticides by soybean looper, Chrysodeixis includens (Lepidoptera: Noctuidae), in Brazil. *Chemosphere* **2020**, *259*, 127499. [CrossRef]
6. Rodrigues-Silva, N.; Canuto, A.F.; Oliveira, D.F.; Teixeira, A.F.; Amaya, O.F.S.; Picanço, M.C.; Pereira, E.J.G. Negative cross-resistance between structurally different Bacillus thuringiensis toxins may favor resistance management of soybean looper in transgenic Bt cultivars. *Sci. Rep.* **2019**, *9*, 199. [CrossRef] [PubMed]
7. Heckel, D.G.; Gahan, L.J.; Baxter, S.W.; Zhao, J.; Shelton, A.M.; Gould, F.; Tabashnik, B.E. The diversity of Bt resistance genes in species of Lepidoptera. *J. Invertebr. Pathol.* **2007**, *95*, 192–197. [CrossRef]
8. Bravo, A.; Gill, S.S.; Soberón, M. Mode of action of Bacillus thuringiensis Cry and Cyt toxins and their potential for insect control. *Toxicon* **2007**, *49*, 423–435. [CrossRef] [PubMed]
9. Tiewsiri, K.; Wang, P. Differential alteration of two aminopeptidases N associated with resistance to Bacillus thuringiensis toxin Cry1Ac in cabbage looper. *Proc. Natl. Acad. Sci. USA* **2011**, *108*, 14037–14042. [CrossRef] [PubMed]
10. Xiao, Y.; Zhang, T.; Liu, C.; Heckel, D.; Li, X.; Tabashnik, B.E.; Wu, K. Mis-splicing of the ABCC2 gene linked with Bt toxin resistance in Helicoverpa armigera. *Sci. Rep.* **2014**, *4*, 6184. [CrossRef]
11. Zhang, S.; Cheng, H.; Gao, Y.; Wang, G.; Liang, G.; Wu, K. Mutation of an aminopeptidase N gene is associated with Helicoverpa armigera resistance to Bacillus thuringiensis Cry1Ac toxin. *Insect Biochem. Mol. Biol.* **2009**, *39*, 421–429. [CrossRef]
12. Jin, T.; Chang, X.; Gatehouse, A.M.R.; Wang, Z.; Edwards, M.G.; He, K. Downregulation and Mutation of a Cadherin Gene Associated with Cry1Ac Resistance in the Asian Corn Borer, Ostrinia furnacalis (Guenée). *Toxins* **2014**, *6*, 2676–2693. [CrossRef]
13. Guo, Z.; Kang, S.; Chen, D.; Wu, Q.; Wang, S.; Xie, W.; Zhu, X.; Baxter, S.W.; Zhou, X.; Jurat-Fuentes, J.L.; et al. MAPK Signaling Pathway Alters Expression of Midgut ALP and ABCC Genes and Causes Resistance to Bacillus thuringiensis Cry1Ac Toxin in Diamondback Moth. *PLoS Genet.* **2015**, *11*, e1005124. [CrossRef] [PubMed]
14. Boaventura, D.; Ulrich, J.; Lueke, B.; Bolzan, A.; Okuma, D.; Gutbrod, O.; Geibel, S.; Zeng, Q.; Dourado, P.M.; Martinelli, S.; et al. Molecular characterization of Cry1F resistance in fall armyworm, Spodoptera frugiperda from Brazil. *Insect Biochem. Mol. Biol.* **2020**, *116*, 103280. [CrossRef]
15. Flagel, L.; Lee, Y.W.; Wanjugi, H.; Swarup, S.; Brown, A.; Wang, J.; Kraft, E.; Greenplate, J.; Simmons, J.; Adams, N.; et al. Mutational disruption of the ABCC2 gene in fall armyworm, Spodoptera frugiperda, confers resistance to the Cry1Fa and Cry1A.105 insecticidal proteins. *Sci. Rep.* **2018**, *8*, 1–11. [CrossRef]
16. Pfaffl, M.W.; Tichopad, A.; Prgomet, C.; Neuvians, T.P. Determination of stable housekeeping genes, differentially regulated target genes and sample integrity: BestKeeper—Excel-based tool using pair-wise correlations. *Biotechnol. Lett.* **2004**, *26*, 509–515. [CrossRef]

17. Andersen, C.L.; Jensen, J.L.; Ørntoft, T.F. Normalization of Real-Time Quantitative Reverse Transcription-PCR Data: A Model-Based Variance Estimation Approach to Identify Genes Suited for Normalization, Applied to Bladder and Colon Cancer Data Sets. *Cancer Res.* **2004**, *64*, 5245–5250. [CrossRef]

18. Panizzi, A.R. History and Contemporary Perspectives of the Integrated Pest Management of Soybean in Brazil. *Neotrop. Èntomol.* **2013**, *42*, 119–127. [CrossRef]

19. Funichello, M.; Fern, J.; Grigolli, o.J.; de Souza, B.H.S.; Boiccedil, A.L.; Junior, A.; Busoli, A.C. Effect of transgenic and non-transgenic cotton cultivars on the development and survival of Pseudoplusia includens (Walker) (Lepidoptera: Noctu-idae). *AJAR* **2013**, *8*, 5424–5428. [CrossRef]

20. Tabashnik, B.E.; Gould, F.; Carriere, Y. Delaying evolution of insect resistance to transgenic crops by decreasing dominance and heritability. *J. Evol. Biol.* **2004**, *17*, 904–912. [CrossRef] [PubMed]

21. Marques, L.H.; Castro, B.A.; Rossetto, J.; Silva, O.A.B.N.; Moscardini, V.F.; Zobiole, L.H.S.; Santos, A.C.; Valverde-Garcia, P.; Babcock, J.M.; Rule, D.M.; et al. Efficacy of Soybean's Event DAS-81419-2 Expressing Cry1F and Cry1Ac to Manage Key Tropical Lepidopteran Pests Under Field Conditions in Brazil. *J. Econ. Èntomol.* **2016**, *109*, 1922–1928. [CrossRef] [PubMed]

22. Tindall, K.V.; Siebert, M.W.; Leonard, B.R.; All, J.; Haile, F.J. Efficacy of Cry1Ac:Cry1F proteins in cotton leaf tissue against fall armyworm, beet armyworm, and soybean looper (Lepidoptera: Noctuidae). *J. Econ. Èntomol.* **2009**, *102*, 1497–1505. [CrossRef]

23. Bel, Y.; Sheets, J.J.; Tan, S.Y.; Narva, K.E.; Escriche, B. Toxicity and Binding Studies of Bacillus thuringiensis Cry1Ac, Cry1F, Cry1C and Cry2A Proteins in the Soybean Pests Anticarsia gemmatalis and Chrysodeixis (Pseudoplusia) includens. *Appl. Environ. Microbiol.* **2017**, *83*, e00326-17. [CrossRef]

24. Jurat-Fuentes, J.L.; Heckel, D.G.; Ferré, J. Mechanisms of Resistance to Insecticidal Proteins from Bacillus thuringiensis. *Annu. Rev. Entomol.* **2021**, *66*, 121–140. [CrossRef] [PubMed]

25. Bustin, S.A.; Benes, V.; Garson, J.A.; Hellemans, J.; Huggett, J.; Kubista, M.; Mueller, R.; Nolan, T.; Pfaffl, M.W.; Shipley, G.L.; et al. The MIQE Guidelines: Minimum Information for Publication of Quantitative Real-Time PCR Experiments. *Clin. Chem.* **2009**, *55*, 611–622. [CrossRef] [PubMed]

26. Vandesompele, J.; De Preter, K.; Pattyn, F.; Poppe, B.; Van Roy, N.; De Paepe, A.; Speleman, F. Accurate normalization of real-time quantitative RT-PCR data by geometric averaging of multiple internal control genes. *Genome Biol.* **2002**, *3*, research0034.1. [CrossRef] [PubMed]

27. Perini, C.R.; Tabuloc, C.A.; Chiu, J.C.; Zalom, F.G.; Stacke, R.F.; Bernardi, O.; Nelson, D.R.; Guedes, J.C. Transcriptome Analysis of Pyrethroid-Resistant Chrysodeixis includens (Lepidoptera: Noctuidae) Reveals Overexpression of Metabolic Detoxification Genes. *J. Econ. Èntomol.* **2021**, *114*, 274–283. [CrossRef] [PubMed]

28. Coates, B.S.; Siegfried, B.D. Linkage of an ABCC transporter to a single QTL that controls Ostrinia nubilalis larval resistance to the Bacillus thuringiensis Cry1Fa toxin. *Insect Biochem. Mol. Biol.* **2015**, *63*, 86–96. [CrossRef] [PubMed]

29. Caccia, S.; Moar, W.J.; Chandrashekhar, J.; Oppert, C.; Anilkumar, K.J.; Jurat-Fuentes, J.L.; Ferré, J. Association of Cry1Ac Toxin Resistance in Helicoverpa zea (Boddie) with Increased Alkaline Phosphatase Levels in the Midgut Lumen. *Appl. Environ. Microbiol.* **2012**, *78*, 5690–5698. [CrossRef]

30. Pigott, C.R.; Ellar, D.J. Role of Receptors in Bacillus thuringiensis Crystal Toxin Activity. *Microbiol. Mol. Biol. Rev.* **2007**, *71*, 255–281. [CrossRef]

Article

Genetic Screening to Identify Candidate Resistance Alleles to Cry1F Corn in Fall Armyworm Using Targeted Sequencing

Katrina Schlum [1], Kurt Lamour [1,2], Peter Tandy [2], Scott J. Emrich [1,3], Caroline Placidi de Bortoli [2], Tejas Rao [2], Diego M. Viteri Dillon [4], Angela M. Linares-Ramirez [5] and Juan Luis Jurat-Fuentes [1,2,*]

[1] Genome Science and Technology Graduate Program, University of Tennessee, Knoxville, TN 37996, USA; kschlum@vols.utk.edu (K.S.); klamour@utk.edu (K.L.); semrich@utk.edu (S.J.E.)
[2] Department of Entomology and Plant Pathology, University of Tennessee, Knoxville, TN 37996, USA; ptandy@vols.utk.edu (P.T.); cplacidi@utk.edu (C.P.d.B.); trao2@utk.edu (T.R.)
[3] Department of Electrical Engineering and Computer Science, University of Tennessee, Knoxville, TN 37996, USA
[4] Isabela Research Substation, Department of Agro-Environmental Sciences, University of Puerto Rico, Isabela, PR 00662, USA; diego.viteri@upr.edu
[5] Lajas Research Substation, Department of Agro-Environmental Sciences, University of Puerto Rico, Lajas, PR 00667, USA; angela.linares@upr.edu
* Correspondence: jurat@utk.edu; Tel.: +1-(865)-974-5931

Citation: Schlum, K.; Lamour, K.; Tandy, P.; Emrich, S.J.; de Bortoli, C.P.; Rao, T.; Viteri Dillon, D.M.; Linares-Ramirez, A.M.; Jurat-Fuentes, J.L. Genetic Screening to Identify Candidate Resistance Alleles to Cry1F Corn in Fall Armyworm Using Targeted Sequencing. *Insects* **2021**, *12*, 618. https://doi.org/10.3390/insects12070618

Academic Editors: Ralf Nauen and Gaelle Le Goff

Received: 26 May 2021
Accepted: 30 June 2021
Published: 8 July 2021

Publisher's Note: MDPI stays neutral with regard to jurisdictional claims in published maps and institutional affiliations.

Simple Summary: Monitoring of resistance alleles is critical to the sustainability of transgenic crops producing insecticidal Cry proteins. Highly sensitive and cost-effective DNA-based methods are needed to improve current bioassay-based resistance screening. Our goal was to evaluate the use of targeted sequencing in detecting known and novel candidate resistance alleles to Cry proteins. As a model, we used field-collected fall armyworm (*Spodoptera frugiperda*) from Puerto Rico, the first location reporting continued practical field-evolved *S. frugiperda* resistance to corn producing the Cry1F insecticidal protein, and sequenced the *SfABCC2* gene previously identified as critical to Cry1F toxicity. Targeted sequencing of *SfABCC2* detected a previously reported Cry1F resistance allele and mutations originally identified in populations from Brazil. Importantly, targeted sequencing also identified nonsynonymous and frameshift mutations as novel candidate resistance alleles. These results advocate for the use of targeted sequencing in screening for resistance alleles to Cry proteins and support potential gene flow, including resistance alleles, between *S. frugiperda* from Brazil and the Caribbean.

Abstract: Evolution of practical resistance is the main threat to the sustainability of transgenic crops producing insecticidal proteins from *Bacillus thuringiensis* (Bt crops). Monitoring of resistance to Cry and Vip3A proteins produced by Bt crops is critical to mitigate the development of resistance. Currently, Cry/Vip3A resistance allele monitoring is based on bioassays with larvae from inbreeding field-collected moths. As an alternative, DNA-based monitoring tools should increase sensitivity and reduce overall costs compared to bioassay-based screening methods. Here, we evaluated targeted sequencing as a method allowing detection of known and novel candidate resistance alleles to Cry proteins. As a model, we sequenced a Cry1F receptor gene (*SfABCC2*) in fall armyworm (*Spodoptera frugiperda*) moths from Puerto Rico, a location reporting continued practical field resistance to Cry1F-producing corn. Targeted sequencing detected a previously reported Cry1F resistance allele (*SfABCC2mut*), in addition to a resistance allele originally described in *S. frugiperda* populations from Brazil. Moreover, targeted sequencing detected mutations in *SfABCC2* as novel candidate resistance alleles. These results support further development of targeted sequencing for monitoring resistance to Bt crops and provide unexpected evidence for common resistance alleles in *S. frugiperda* from Brazil and Puerto Rico.

Keywords: *Spodoptera frugiperda*; fall armyworm; resistance; Cry1F; genotyping; targeted sequencing; resistance screen; ABCC2

1. Introduction

The fall armyworm (*Spodoptera frugiperda*, J.E. Smith) (Lepidoptera: Noctuidae) is a global invasive pest affecting numerous food and fiber staple crops, although the highest damage is observed in corn (*Zea mays* L.) [1]. In North America, *S. frugiperda* moths display long-distance northward migration over several generations from overwintering sites in southern Texas and Florida, reaching Canada in the summer months [2]. Substantial gene flow also occurs between *S. frugiperda* populations from the Caribbean and Florida, but not between these populations and moths from South America [3].

Effective control of *S. frugiperda* in the Western hemisphere in addition to pesticides has been provided by genetically modified corn and cotton producing insecticidal proteins from the bacterium *Bacillus thuringiensis* (*Bt*), namely Cry1F, Cry1Ab, Cry1A.105, Cry2Ab and Vip3Aa. In the past decade, resistance to Cry1F, Cry1Ab and Cry1A.105 was reported for *S. frugiperda* populations in Puerto Rico and the continental US (Florida and North Carolina) [4,5], Brazil [6] and Argentina [7]. Fall armyworm resistance to these Cry proteins is recessive and genetically linked to mutations in an ABC transporter superfamily C2 gene (*SfABCC2*) [8–10]. In Puerto Rico, two *SfABCC2* resistance alleles have been reported: a nonsense 2 bp insertion (*SfABCC2mut* allele) [8] and an insertion near the start of the fourth exon associated with aberrant splicing [9]. In contrast, resistance to Cry1F in Brazil is linked to missense mutations primarily localized to the fourth extracellular loop in SfABCC2, in support of this region containing a Cry1F binding domain [11] and being a hotspot for resistance mutations [10]. Genetic knockout of *SfABCC2* leads to resistance against Cry1F [12,13] further supporting that truncation of SfABCC2 is predictive of a Cry1F-resistant phenotype.

Identification of resistance alleles in *S. frugiperda* is critical to developing highly sensitive DNA-based genotyping tools, which are less laborious and costly compared to F1 [14] or F2 [5] resistance screens. Successful *SfABCC2* genotyping efforts were described using TaqMan probes [8,9] and pyrosequencing [10] in populations from Puerto Rico, USA and Brazil. Results from these tests support high-resistance allele frequencies in areas where practical resistance had been reported, further associating identified *SfABCC2* alleles with field-relevant resistance. An unexpected observation, considering predicted migratory models [3], was not detecting the *SfABCC2mut* allele reported in Puerto Rico at migratory Caribbean destinations or in the Florida populations tested [8,9]. Relevant fitness costs are not associated with *SfABCC2mut* under laboratory conditions [15] and thus probably do not explain the limited distribution to Puerto Rico of *SfABCC2mut*.

Hi-Plex (HP) is a highly multiplexed PCR-based approach for targeted massive parallel sequencing with demonstrated efficacy in disease gene screening [16,17]. It involves amplifying small (e.g., 180 bp) overlapping portions of a larger target region (tiling) in one (or a few) multiplex PCR reaction(s) adding a unique sample-specific barcode sequence during the PCR reaction. The HP technology is attractive because it requires very little genomic DNA to amplify multiple target amplicons (e.g., 20–40 ng of template DNA to amplify a few hundred targets), the DNA can be highly complex (e.g., mixed with genomic DNA from many different organisms) and the overall cost in time and reagents is greatly reduced compared to F1 and F2 screens. The barcoded amplicons are pooled and sequenced on a next-generation sequencing (NGS) device, and the sample-specific amplicons mapped to a reference genome.

The goal of this project was to adapt and provide preliminary data for a new HP targeted screen for resistance alleles to *Bt* insecticidal proteins. We focused our analysis on *S. frugiperda* samples from Puerto Rico, as both *SfABCC2mut* and Cry1F resistance are highly frequent at this location [8,18]. Furthermore, mutations resulting in a truncated SfABCC2 protein result in a Cry1F-resistant phenotype [8,9,12,13], potentially allowing for phenotypic predictions. Analysis of Taqman and targeted *SfABCC2* sequences detected the known *SfABCC2mut* allele in 34–37% of tested samples, depending on the technique used. In addition, the HP approach coupled to a bioinformatics pipeline detected one nonsense and nine frameshift *SfABCC2* mutations as novel candidate Cry1F resistance alleles. Missense mutations (13) were also identified, but phenotype predictions are not possible in this case as the effect of these mutations on Cry1F susceptibility and resistance requires functional testing. We document, for the first time, that a *SfABCC2* resistance allele previously described in *S. frugiperda* from Brazil is present in Puerto Rico fall armyworm populations. These data provide proof of concept for the power of HP targeted sequencing for performing surveillance and field monitoring of well-established and novel candidate *Bt* resistance alleles.

2. Materials and Methods

2.1. S. frugiperda Sample Collection and Genomic DNA Extraction

Adult moths were captured at three locations in Puerto Rico (Lajas, Salinas and Santa Isabel) using sex pheromone-baited traps [19] at sites near field and sweet/field corn plantings in order to optimize the capture efficiency of C-strain males. Collection details of the 41 *S. frugiperda* individuals tested are in Table S1. Captured moths were confirmed as *S. frugiperda* by visual inspection [20] and kept refrigerated until shipped to the University of Tennessee. Samples were stored at −20 °C until processed for purification of genomic DNA. Samples were named and numbered according to trapping location as PRL (Lajas), PRS (Salinas) and PRSI (Santa Isabel). A sample from the previously characterized Cry1F-resistant 456LSD4 strain of *S. frugiperda*, which originated from collections in Puerto Rico [21], was also included as positive control for detection of *SfABCC2mut* [8].

Genomic DNA was isolated from moth legs using the Pure Link Genomic DNA mini kit (Invitrogen) following manufacturer's protocols, and then quantified using a Nanodrop spectrophotometer (Thermo Scientific).

2.2. TaqMan Genotyping for SfABCC2mut

Detection of the *SfABCC2mut* allele using Taqman probes was performed as previously described [8]. Briefly, 10 µL (final volume) reactions included 10–20 ng of gDNA as template, a VIC-labeled probe specific to the *SfABCC2mut* allele (5′ AAGCACATCGCCCACTT 3′), a FAM-labeled probe specific to the wild-type *SfABCC2* allele (5′ CCAAGCACATCCCACTT 3′), and forward (5′ TGGAGGCCGAAGAGAGACA 3′) and reverse (5′ AGGAGTTGACT-GACTTCATGTACCT 3′) primers. Plates (Micro Amp Fast optical 96 well reaction plate, Applied Biosystems) were loaded in the Quant studio 6 Real Time PCR instrument (Applied Biosystems) and amplified as follows: pre read stage at 60 °C for 30 s, hold stage at 95 °C for 10 min, PCR stage at 95 °C for 15 s and 60 °C for 1 min for 40 cycles, post read stage at 60 °C for 30 s. The fluorescence in each well was measured in the post read stage of the PCR, and the software generated an allelic discrimination plot based on the post amplification intensity of the fluorescent probes.

The frequency of the *SfABCC2mut1* allele or any mutation was determined using the Hardy–Weinberg equation with the formula: F = (2 × ObsAa + Obsaa)/[2 × (ObsAA + ObsAa + Obsaa)], where "F" is the frequency of the "a" allele (*SfABCC2mut1*) and "Obs" the observed frequency of each of the three possible genotypes for the allele.

2.3. Targeted DNA Amplification and Sequencing

In early BLASTn searches of the *S. frugiperda* corn strain v3.1 genome from LepidoDB (https://bipaa.genouest.org/sp/spodoptera_frugiperda_pub/, accessed on 11 July 2019), we found that the *SfABCC2* gene was split between Scaffolds 11087 and 7154. These scaffolds were manually combined and the resulting *SfABCC2* gene sequence was used as a template to design TILING primers, using PCRTiler [22] at default settings allowing 50 bp overlap on each end and producing 180–200 bp final amplicon sizes (Table S2). The two approximately 1 Kbp intronic regions at the beginning of that gene sequence (Scaffold_11087) were not targeted. In a later version of the *S. frugiperda* corn strain genome (v6.0) from LepidoDB (accessed on 16 November 2020), the full-length *SfABCC2* gene appears in Scaffold 33. The initial *SfABCC2* gene assembly based on Scaffolds 11087 and 7154 used for TILING primer design did not include an inter-scaffold gap encompassing nucleotides 7239–8976 of the *SfABCC2* gene assembly in the v6.0 genome. The TILING primers covered approximately 90% of the exonic regions of the assembled *SfABCC2* gene found in Scaffold 33 of the *S. frugiperda* corn strain v6.0 genome (see Figure S1 for a visual representation of typical coverage).

All TILING primers (238) were mixed into a single multiplex (119 overlapping targets) and PCR performed using the Hi-Plex targeted sequencing strategy, as previously described [17]. The resulting amplicons were pooled and sequenced on a HiSeqX device running a 2 × 150 bp configuration and merged to produce single, high quality reads using PEAR [23]. Demultiplexing was accomplished using the sample-specific barcode sequence incorporated during the Hi-Plex PCR amplification.

2.4. Genotype Assignment for SfABCC2mut Using a Targeted-Sequencing k-mer-Based Approach

To genotype the *SfABCC2mut* 2 bp insertion in the *SfABCC2* targeted sequencing data, a custom Python 3 script (dubbed Pgeno) was developed and is deposited in Github at https://github.com/petertandy/pgeno (accessed on 26 May 2021). This script scans each of the sequence reads for the known mutation by using 10 mer flanking sequences as a guide. To account for unknown variation in the flanking regions, each of the 10 mers was allowed up to two mismatches. Sequences with matching flanks were counted and sample assigned a *SfABCC2mut* genotype based on the proportion of the known 2 bp insertion. In order to reduce false-positive genotype assignment, we required at least 40X sequence coverage in the PCR targeted regions and >15% of the total number of matching reads to assign a genotype of hetero- or homozygous for the 2 bp insertion. The minimum sequence coverage refers to the number of high-quality reads covering a target region in the *SfABCC2* reference gene and depends on the efficiency of the primers in the multiplex. The primer efficiencies are highly reproducible and by over-sequencing the samples, most of the gene is covered at a high depth (up to 1000X for some targets). These parameters can be adjusted in the Pgeno script to increase or decrease the genotyping stringency.

2.5. Alignment, Variant Calling and Variant Analysis

The complete pipeline used for trimming, alignment, variant calling, and analysis using targeted *SfABCC2* sequences is summarized in Figure 1. Raw reads were quality trimmed at both ends and filtered for adapter sequences using BBDuk [24]. The quality of samples was checked before and after trimming/processing to confirm adapter and non-relevant (e.g., contaminants, primer artifacts) sequences were removed using FastQC [25]. The full-length *SfABCC2* gene interval was determined by aligning a *SfABCC2* cDNA from a Cry1F-susceptible *S. frugiperda* strain (GenBank accession number KY489760.1) to the *S. frugiperda* corn reference genome v6.0 [26] via the BLAST server hosted on LepidoDB. The full-length *SfABCC2* gene sequence was found included in Scaffold 33, and this reference gene sequence was used for further analyses.

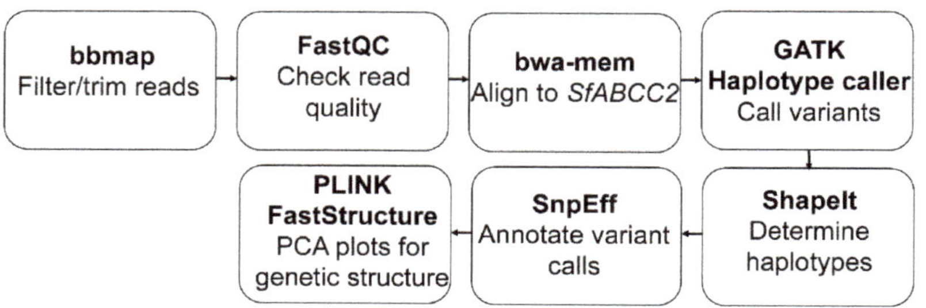

Figure 1. Schematic pipeline of overall workflow used in this work from raw Fastq targeted sequencing reads to population genetics and phylogenetic analyses based on bi-allelic SNP variant calling.

Reads were aligned to the *SfABCC2* reference gene using a Burrows–Wheeler aligner algorithm (bwa-mem) with default options [27]. The mapping rate for all samples was determined via SAMtools flagstat [28]. Only samples with a mapping rate of 80% or higher were kept, which led to the removal of only three samples from further analysis. We determined the count of DNA sequence reads across the *SfABCC2* gene using bam-readcount (https://github.com/genome/bam-readcount, accessed on 20 June 2019). As noted in 2.3, two regions (nucleotides ~138–1518 and ~7195–7858) within the *SfABCC2* gene were not sequenced due to the TILING primers being designed based on an old assembly (Scaffolds 11,087 and 7154 from *S. frugiperda* corn genome v3.1) of the gene. The second region representing the inter-scaffold gap in the early gene assembly was filled with 'Ns', precluding alignment of any reads to this region.

We performed alignment cleanup by sorting the bam file via SAMtools [28], deduplicating via Genome Analysis Tool Kit (GATK) version 4.1.20 [29] with MarkDuplicates and then adding read groups to each bam file via GATK's function addOrReplaceReadGroup. Variants were called via the Haplotypecaller of GATK with default parameters set based on the GATK Best Practice pipeline. The initial set included 2006 variants with a quality score greater than 40. After considering different filters based on depth and minor allele frequencies, the final filter applied included depth of coverage ≥ 8, each individual allele depth ≥ 2, alternate allele frequency $\geq 5\%$, and the allele being called in at least 25% of the samples (>10). This filter produced 1861 variants (1428 SNPs and 433 indels), over 92% of the unfiltered set. We further filtered this variant set to include only bi-allelic variants using the GATK SelectVariants function, which resulted in 1333 bi-allelic variants (1143 SNPs and 190 indels) that were used in all variant-based analysis.

The intron/exon boundaries in *SfABCC2* were determined by alignment to a *SfABCC2* cDNA (GenBank accession number KY489760.1) using Exonerate with the est2genome model option [30]. To confirm the accuracy of the intron/exon boundaries, we performed the same alignment of the *SfABCC2* gene with three additional available cDNA sequences from Cry1F-susceptible *S. frugiperda* (GenBank accession numbers MG387043.1, MN399979.1 and KY646296.1). The same intron/exon boundaries were identified in each generic feature file (GFF) generated by Exonerate. The consensus protein sequence from aligning the SfABCC2 sequences derived from translating the *SfABCC2* cDNA sequences above in CLC Genomics Workbench 21.0.4 (Figure S2) was used as a reference to locate amino acid changes from mutations. This consensus SfABCC2 protein sequence differs in two amino acid positions in the notation for mutations located to extracellular loop 4 region presented by Boaventura et al. [10].

Functional annotation was performed with SnpEff [31], and SnpSift was used to filter annotations to a tabular format [32]. The final filtered set of 1333 variants contained 17 missense mutations and 1 nonsense mutation. Compared to the unfiltered variant set, the filtered set excluded one nonsense mutation because of both alternate allele frequency and read depth criteria, and two missense mutations because of the alternate allele frequency criterion.

We also computed PCA projections via PLINK [33], and visualized them in ggplot2 [34]. Given the highly polymorphic nature of the samples, we also determined haplotypes for PCA and generated neighbor joining trees using the ShapeIT software [35]. Since no recombination maps are available for *S. frugiperda*, the recombination rate in ShapeIT was set to 0.000023 cM/Mb based on previous research for *Spodoptera litura* [36].

Admixture model-based clustering was performed using FastStructure [37] to identify genetic structure within the *SfABCC2* gene. FastStructure is a generative model-based approach based on Hardy–Weinberg equilibrium assumptions between alleles and linkage disequilibrium between genotyped loci. The FastStructure script structure.py was used to determine k, the number of assumed populations or genetic groups that share a subset of allele frequencies from $k = 2$ to 10. The choosek.py script in FastStructure chose $k = 3$ for model complexity maximizing the marginal likelihood, and $k = 4$ for model components to explain structure in the data. The final output of FastStructure was visualized using distruct.py at $k = 4$.

3. Results

3.1. Genotyping Results Using Taqman

Based on Taqman genotyping, of the 41 *S. frugiperda* samples tested, 18 carried the *SfABCC2mut* allele, with nine being homozygous. Overall, these detections suggest an allele frequency of 0.3292 for *SfABCC2mut* in the analyzed samples, similar to previous estimations for the frequency of this allele in Puerto Rico [8]. The relative proportion of individuals carrying the *SfABCC2mut* allele at each location was higher in Lajas (53% carriers, including 33% homozygotes and 20% heterozygotes) followed by the Salinas (41% carriers, 8% homozygotes and 33% heterozygotes) and Santa Isabel (21% homozygotes) locations.

3.2. Genotyping Results Using Targeted Sequencing

Overall, the multiplexed TILING primers amplified 63% of the *SfABCC2* reference gene from the *S. frugiperda* v.6.0 corn genome, with 23 of the 25 predicted exons located in contig 33 being fully sequenced (Figure S1). The Hi-Plex strategy does not rely on equilibration of primer efficiencies (or any kind of primer optimization) and as such, some targets will amplify more efficiently than others (or fail completely) and thus receive more sequencing (or none at all). To accommodate these expected variations in template concentration, the samples were sequenced to high depth ($>100\times$ per target amplicon).

Of the 41 samples genotyped via targeted sequencing, 14 contain the *SfABCC2mut* GC insertion, with 8 scored as heterozygous and 6 as homozygous, resulting in a 0.2683 allele frequency. There was a 73% (30/41) concordance rate between the Taqman and the k-mer-based *SfABCC2mut* genotyping (Table 1). Five samples identified as susceptible by the k-mer-based genotyping were identified as heterozygotes (3) or homozygote resistant (2) by the Taqman assay. Furthermore, three heterozygous samples, based on k-mer-based genotyping, appeared as homozygous resistant in Taqman tests. Finally, two samples identified as homozygous resistant by the k-mer method were classified as homozygous susceptible by the Taqman assay. Examination of the assembled targeted sequences for the discrepant samples detected a number of SNPs in the *SfABCC2mut* site targeted by Taqman primers, which likely affected Taqman amplifications and therefore explain the observed discrepancies. In fact, of the 58 sites covered by the Taqman primers and probe, 19% were polymorphic in at least one of the 41 samples (data not shown).

3.3. Variant Analysis Pipeline and Genotyping

Functional annotation of variants detected using the bioinformatics pipeline shown in Figure 1 detected *SfABCC2mut* (2 bp GC insertion) as a frameshift mutation (D740A) There was a total of 18 samples in which the *SfABCC2mut* allele was detected using the GATK variant and filtering described earlier (see Methods). Within these 18 samples, 12 were heterozygous and 6 homozygous for the mutation, displaying 100% concordance with results from the k-mer-based genotyping approach. When a lower alternative allele threshold (≥ 3) was utilized, some discrepancies occurred between GATK and the k-mer-based genotyping (data not shown).

In addition to identifying the known *SfABCC2mut* allele, we also detected 374 coding synonymous mutations (data not shown) and 23 nonsynonymous and frameshift mutations in *SfABCC2* in the filtered dataset. After mutations already observed in at least one *SfABCC2* cDNA from Cry1F-susceptible *S. frugiperda* were removed, 14 nonsynonymous and 9 frameshift mutations (apart from *SfABCC2mut* being synonymous with D740A) were identified as new potential Cry1F resistance alleles (Table 2 and Table S1).

Table 1. Concordance between results using Taqman and k-mer-based genotyping for the *SfABCC2mut* (GC insertion) allele, represented as the "r" allele. Samples with discrepant genotypes are highlighted in bold font.

Sample Name	Taqman	K-mer
PRL_164	SS	SS
PRL_165	**rr**	**Sr**
PRL_166	SS	SS
PRL_167	SS	SS
PRL_168	Sr	Sr
PRL_169	**rr**	**Sr**
PRL_170	Sr	Sr
PRL_171	Sr	Sr
PRL_172	rr	rr
PRL_173	rr	rr
PRL_174	SS	SS
PRL_175	SS	SS
PRL_176	SS	SS
PRL_177	rr	rr
PRL_178	**SS**	**SS**
PRS_6	**SS**	**SS**
PRS_7	Sr	Sr
PRS_8	SS	SS
PRS_9	**Sr**	**SS**
PRS_10	rr	rr
PRS_11	SS	SS
PRS_12	**Sr**	**SS**
PRS_13	SS	SS
PRS_14	SS	SS
PRS_17	**Sr**	**SS**
PRS_18	**SS**	**rr**
PRS_20	SS	SS
PRSI_1	SS	SS
PRSI_2	SS	SS
PRSI_4	SS	SS
PRSI_5	**SS**	**Sr**
PRSI_6	**rr**	**Sr**
PRSI_7	**rr**	**SS**
PRSI_8	SS	SS
PRSI_9	**SS**	**rr**
PRSI_10	**rr**	**SS**
PRSI_11	SS	SS
PRSI_12	SS	SS
PRSI_13	SS	SS
PRSI_14	SS	SS
PRSI_15	SS	SS

All missense (13) and nonsense (1) nonsynonymous mutations, except for an homozygous sample for T14M, were detected in heterozygous individuals. The most abundant missense mutations (10 individuals each, 24% of samples) were G1085E and S1209N, which occur in the C terminal intracellular region of SfABCC2 (Figure 2). The other common missense mutation was V414A (carried by 9 individuals, 22% of samples), which occurs in the loop connecting the first SfABCC2 transmembrane domain and ATP-binding cassette. Mutation P1079L was localized to a predicted splice site region. Only one missense mutation (Q788P, found in 1 sample) was localized to the fourth extracellular loop, a region hypothesized to be vital to Cry1F toxicity [10,11]. Missense mutations had a predicted moderate effect on the SfABCC2 protein, while the only nonsense mutation (E700 *) was predicted to result in a truncated SfABCC2 protein encompassing up to the first ATP-binding cassette (Figure 2).

Table 2. Nonsynonymous and frameshift mutations identified by SnpEFF across 41 *S. frugiperda* *SfABCC2* bi-allelic variants. The nucleotide position in the *SfABCC2* gene reference (Position), the functional class of mutation (MISS = missense, NON = nonsense, and SHIFT = frameshift), the mutation with capitalized reference and alternative alleles, the corresponding amino acid change (Aa change), and the number of samples positive for that mutation (#Samples) are shown. Asterisks denote that the mutation leads to a premature stop codon. The *SfABCC2mut* allele is synonymous with D740A.

Position	Class	Mutation	Aa Change	#Samples
38	SHIFT	ggc/ggTGGTc	G13GG	1
40	SHIFT	acg/aATGGTGGTCATcg	T14NGGH *	1
41	MISS	aCg/aTg	T14M	4
42	SHIFT	gct/	A15 *	20
45	SHIFT	gtg/gGTCAACG	V16GQR *	4
48	SHIFT	ggc/CAggc	G17Q *	4
3421	SHIFT	gtc/gCATGtc	V210AC *	1
4077	MISS	Ctc/Ttc	L330F	2
4357	MISS	Ttt/Gtt	F398V	1
4540	MISS	gTg/gCg	V414A	9
6020	MISS	cAa/cGa	Q627R	1
6915	NON	Gaa/Taa	E700 *	1
7034	SHIFT	gat/gCGat	D740A *	18
7070	SHIFT	tgcatggtg/	CMV753	1
7120	MISS	gCt/gGt	A768G	3
9098	MISS	cAa/cCa	Q788P	1
9122	SHIFT	gga/GTAAgga	G793VR	8
9335	MISS	gCc/gTc	A822V	3
10491	SHIFT	gga/	G950	1
11708	MISS	cCg/cTg	P1079L	1
11726	MISS	gGa/gAa	G1085E	10
12507	MISS	Gaa/Aaa	E1195K	2
12748	MISS	aGt/aAt	S1209N	10
13285	MISS	Tcc/Acc	S1310T	1

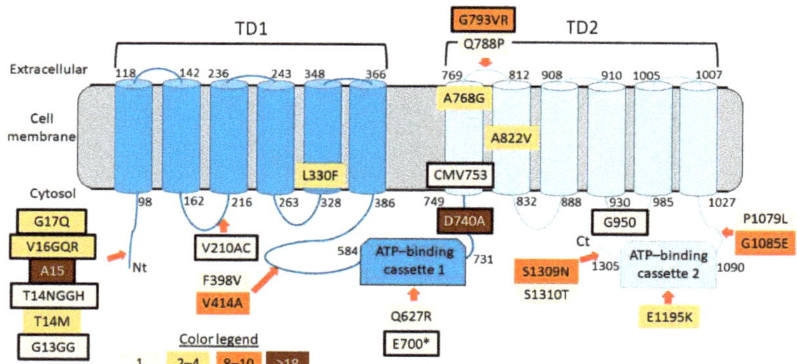

Figure 2. Schematic representation (not drawn to scale) of SfABCC2 topology on the cell membrane and location of detected nonsynonymous, nonsense and frameshift mutations in sequenced S. frugiperda samples from Puerto Rico. Amino acid numbers are indicated at the start and end of transmembrane domains and ATP-binding cassettes as predicted based on detection of conserved domains in BLAST searches. Distinct colors in boxes indicate the number of individuals carrying each mutation, as detailed in the color legend. The only nonsense mutation (E700 *) and all frameshift mutations are shown in a boxes with bold borders, including the *SfABCC2mut* allele (D740A). TD = transmembrane domain.

Frameshifts were comparatively more frequent than nonsynonymous mutations, with homozygous individuals detected for all the frameshift mutations for which more than one positive sample was detected. These frameshift mutations generated aberrant proteins and predicted premature stop codons, leading to truncated SfABCC2 proteins, which in homozygous individuals would correspond to a Cry1F-resistant phenotype [8,12]. The A15 mutation was the most common (49% of individuals), with a slightly lower allele frequency (0.3170) than *SfABCC2mut* (synonymous with D740A mutation). Only three samples (PRL_168, PRL_170, PRL_171) from the same location (Lajas) were heterozygous for both A15 and D740A. These samples could be resistant to Cry1F based on complementation of these two alleles.

3.4. SfABCC2 Sequence Variation

We also considered population-level diversity in *SfABCC2* sequences from Puerto Rico. Interestingly, a Principal Component Analysis (PCA) of all samples showed two distinct groups, one of them without *SfABCC2mut* in any individual (Figure 3A, right). Individuals homozygous for *SfABCC2mut* had the most compact cluster, suggesting more *SfABCC2* sequence variation within the SS compared to rr samples.

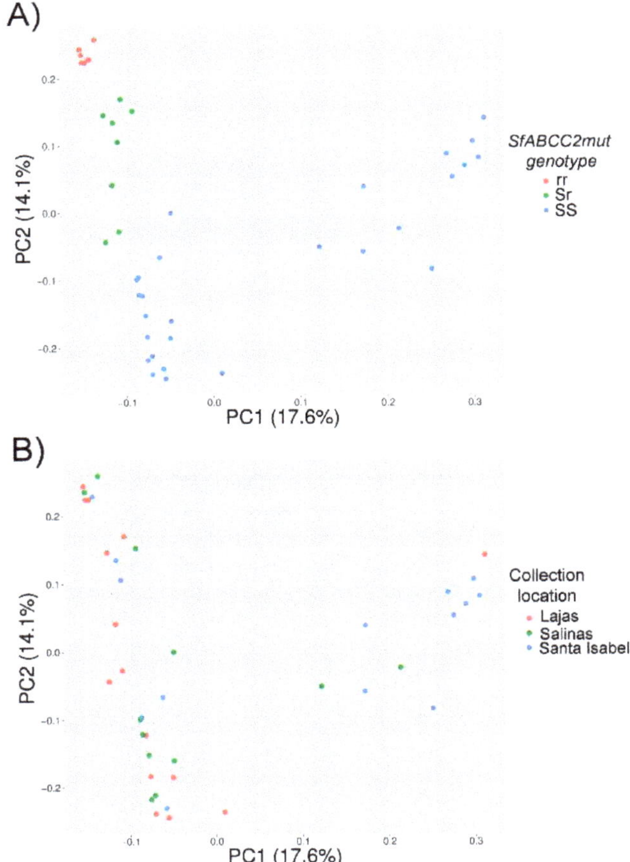

Figure 3. PCA projection plot of 1333 bi-allelic variants found within *SfABCC2* of 41 *S. frugiperda* samples and colored based on genotype for *SfABCC2mut* (**A**) and location (**B**). The *SfABCC2mut* genotype was determined as described in the Methods.

To determine whether location was driving differentiation of the PC clusters observed given different frequencies of *SfABCC2mut* in the three Puerto Rico locations, we also plotted these samples colored by location (Figure 3B). No clusters correlated with sample location. Further, a logistic regression association test on PC1 vs. PC2 (Figure 4) generated eight SNPs with significant -log 10 *p*-value > 5 at positions 3874, 3936, 3944, 4085, 4097, 5485, 5788, and 13206. None of these variants, however, were predicted to have any functional effect change by SnpEff or corresponded to previously reported Cry1F-resistant alleles [8,10]. These data suggest that there may be barriers to Bt resistance gene flow, for example between host strains, even within the island of Puerto Rico.

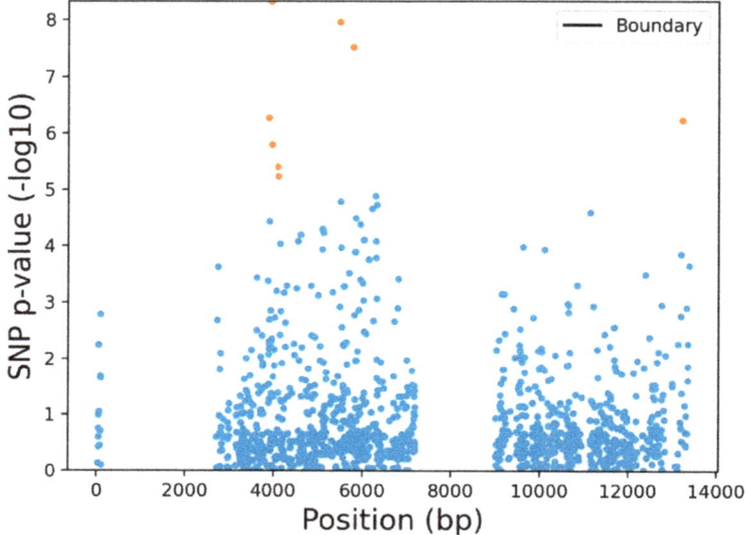

Figure 4. Manhattan plot of SNPs associated with PC2 (differentiating clusters of Sr and rr genotyped samples) on the filtered bi-allelic dataset (1333 variants).

To confirm this observation, we re-ran the PCA analysis without the *SfABCC2mut* homozygous individuals and observed the same clusters (data not shown). We next haplified our *SfABCC2* data using ShapeIt2 (1333 variants, see Methods) and ran FastStructure (Figure 5). Based on the targeted sequence data, we observed four populations, with the *SfABCC2mut* allele always associating with one population (purple color in Figure 5). Furthermore, even though there is little observed sharing of *SfABCC2* between the two primary PCA-based groups, admixture did occur between the *SfABCC2mut*-associated population and the three other populations. The frequency of the *SfABCC2mut* allele was highest in Lajas, indicating that this allele likely originated from this location and spread to other fall armyworm populations in Puerto Rico, as seen in the Structure plot. The observed moderate levels of admixture between the fall armyworms studied across all three Puerto Rican populations suggest that—contrary to the PCA results—the *SfABCC2mut* has the potential to spread to other subpopulations.

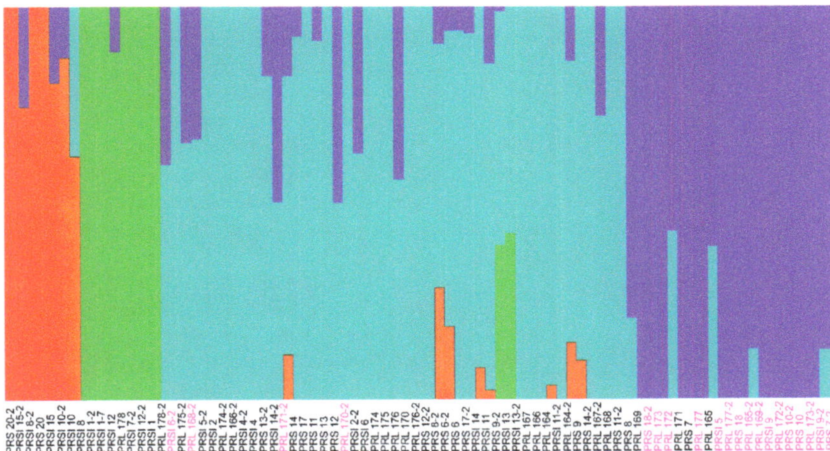

Figure 5. Genetic structure inferred by FastStructure on 82 haplotypes representing 41 samples at $k = 4$. Samples in pink text contain the *SfABCC2mut* allele.

4. Discussion

Bioassay-based testing of F1 or F2 generations has traditionally been used to screen for resistance alleles in *S. frugiperda* to insecticidal proteins from *B. thuringiensis* produced in transgenic crops [5,14,38,39]. As an alternative, DNA-based screening methods provide higher sensitivity and are amenable to high-throughput applications and can considerably reduce overall labor and costs. Initial DNA-based resistance monitoring efforts for *S. frugiperda* focused on Taqman PCR targeting known mutations in the *SfABCC2* gene [8,9] and pooled population sequencing [10]. More recently, whole-genome resequencing was used to detect the presence of five reported resistance alleles in *S. frugiperda* populations from 12 geographic locations [40]. As a more cost-effective alternative to whole-genome sequencing, we provide proof of concept for the use of targeted highly multiplexed PCR for monitoring a known resistance allele (*SfABCC2mut*) and using bioinformatics to identify novel candidate *SfABCC2* resistance alleles. An advantage of the Cry1F-SfABCC2 system is that available knockout [12,13], mutational [11] and resistance [8,9] data in *S. frugiperda* support that a truncated SfABCC2 protein cannot function as Cry1F receptor. Thus, homozygote individuals for nonsense or frameshift mutations leading to truncated SfABCC2 proteins display a Cry1F-resistant phenotype, as observed for the *SfABCC2mut* resistance allele [8,9]. In addition, the amplification strategy of Hi-Plex allows the use of highly fragmented DNA, suggesting that poorly preserved samples, as is often the case for adult moths collected with pheromone traps, are amenable to analysis.

Both a k-mer-based and a bioinformatics pipeline provided the same genotyping results for a known mutation (*SfABCC2mut*). The genotypes of some samples differed between the Hi-Plex and Taqman methods. We observed that 19% of tested samples had polymorphisms around the priming/probe site for Taqman, which suggests an advantage of Hi-Plex over Taqman when dealing with highly polymorphic resistance genes. Most likely, the overlapping targets (and thus different priming sites) used in the TILING approach helps reduce the potential for erroneous genotypes and conclusions.

We found that among the individuals carrying the *SfABCC2mut* allele, 14.6% were homozygotes, which are expected to produce a truncated SfABCC2 and thus be resistant to Cry1F [8,9,12,13]. The allele frequency of *SfABCC2mut* (0.3292) was in line with previous screening efforts in fall armyworm from Puerto Rico [8,9]. Based on the Faststructure and PCA-based analysis, no clear selective sweep was observed. The higher relative frequency of *SfABCC2mut* suggests the allele may have emerged in *S. frugiperda* in or around Lajas. Our population-level analysis suggests that the genomic region around *SfABCC2mut* is subject to both recombination and introgression, and that this occurs frequently, especially between populations at Santa Isabel and Lajas. In contrast, we still observed two main populations in Puerto Rico with statistically significant markers that do not associate with known or candidate resistance alleles. Further work is needed to determine whether there is at least one barrier to large-scale gene flow on the island, e.g., the well-established Corn vs. Rice strains in *S. frugiperda*. It is important to consider that Lajas has had rice winter nurseries (Univ. of Puerto Rico and Ricetech) for more than 30 years, while corn is the most prevalent crop at Santa Isabel. Furthermore, organic corn is planted in Lajas, which requires the frequent use of *B. thuringiensis* insecticides (e.g., Dipel) to control Lepidopteran pests [41].

In addition to *SfABCC2mut*, a second Cry1F-resistant allele was previously described in *S. frugiperda* from Puerto Rico [9]. In this case, insertion of a highly repetitive sequence at the beginning of the *SfABCC2* gene resulted in aberrant splicing and resistance. This type of allele would not be detected in our current Hi-Seq strategy as our TILING primers were not designed to amplify the highly repetitive sequence. Addition of primers amplifying this region and re-sequencing may allow detection and estimation of frequency for this allele. However, the reported highly repetitive nature of the insert [9] may hinder successful detection.

Targeted sequencing combined with a custom bioinformatics pipeline allowed detecting, in addition to 374 coding synonymous mutations, 23 nonsynonymous and frameshift *SfABCC2* mutations as novel candidate Cry1F resistance alleles. A Cry1F-resistant phenotype similar to homozygous individuals for *SfABCC2mut* would be predicted for homozygous frameshift and nonsense mutations, as they would result in lack of a functional SfABCC2 protein [9,12,13]. Interestingly, some of the frameshift mutations were noticeably frequent. Importantly, a number of individuals positive for these frameshift mutations were homozygous samples predicted to produce truncated SfABCC2 proteins and thus having a Cry1F-resistant phenotype similar to *SfABCC2mut* [8]. In addition, heterozygous individuals could also be phenotypically resistant to Cry1F in cases of complementation between alleles involving nonsense and frameshift mutations.

Significant gene flow between *S. frugiperda* populations from Brazil and Puerto Rico is not expected based on haplotype marker and meteorological observations [3]. In agreement with this observation, previous genotyping efforts did not detect *SfABCC2mut* among Brazilian *S. frugiperda* populations [9]. However, whole-genome comparisons suggest panmixia among *S. frugiperda* populations [42]. We did find two heterozygote individuals from Lajas (PRL_168 and PRL_164) carrying a GY deletion (DGY784D in our model, allele frequency 0.0571) previously described in Cry1F-resistant and field-collected *S. frugiperda* from Brazil [10]. However, detection of this GY deletion in Cry1F-suceptible individuals from Brazil [40] questions the relevance of this mutation to resistance.

The extracellular loops 1 and 4 in SfABCC2 have been proposed as relevant to Cry1F toxicity [11], with loop 4 being most critical and also a hot spot for resistance mutations [10]. Only the Q788P mutation localized to that loop in our dataset. Interestingly, samples containing this mutation were detected in pooled sequencing of Brazilian *S. frugiperda* populations in one sample each from four locations (Balsas, Dourados, Santa Cruz das Palmeiras, and Uberlandia) [10]. These particular collection sites in Brazil are separated by >1200 km, suggesting that the Q788P mutation may be common in Brazilian *S. frugiperda* populations. Among our samples, the mutation was rare and a single heterozygote individual (PRS_14) was detected. Functional tests would be needed to test the effect of Q788P on Cry1F toxicity.

Detection of common mutations may be suggestive of gene flow, including resistance alleles, between *S. frugiperda* populations from Brazil and the Caribbean. Although this flow is not expected based on predicted migratory movements [3], commercial exchange of contaminated plant materials could result in unintended dissemination of resistance alleles. This hypothesis should be further tested with increased sample sizes to better understand the potential paths for the spread of *S. frugiperda* resistance to *Bt* crops.

Overall, our analyses detected 374 synonymous and 24 (including *SfABCC2mut*) non-synonymous or frameshift *SfABCC2* mutations in 41 tested *S. frugiperda* individuals from Puerto Rico. These results indicate the *SfABCC2mut* and novel mutations predicted to result in truncated or aberrant SfABCC2 proteins and thus a Cry1F-resistant phenotype are not rare in *S. frugiperda* from Puerto Rico. Out of the 41 samples tested, the majority (38, 93%) carried at least one copy of a nonsense or frameshift *SfABCC2* mutation. There were 13 (32% of the total) homozygotes for these mutations, including 6 for *SfABCC2mut* and 6 for the A15 allele. Based on similar mutations in knockout and resistant strains of *S. frugiperda* [8,9,11,13], these homozygotes would display a Cry1F-resistant phenotype. The relatively high frequency of mutations predicted to result in resistance to Cry1F through SfABCC2 truncation is in line with the high levels of resistance to Cry1F maintained in Puerto Rico [18,43] and the lack of relevant fitness costs in disruptive *SfABCC2* mutants [15,44]. Our targeted approach does not allow us to identify additional Cry1F-resistant individuals carrying alternative resistance genes, which could potentially increase the frequency of Cry1F-resistant individuals. In addition, while missense *SfABCC2* mutations were also detected, a Cry1F-resistant phenotype cannot be predicted as their role in intoxication is unknown. For instance, the Q788P and A768G mutations could affect the predicted Cry1F-binding ability of the SfABCC2 loop 4 region [10]. Functional assays would be needed to test the effect of these mutations on susceptibility to Cry1F.

5. Conclusions

We provide proof of concept for the use of targeted sequencing in the monitoring of known mutations and the discovery of candidate resistance alleles to Cry1F in *S. frugiperda*. Furthermore, based on solid experimental evidence for the critical role of SfABCC2 in susceptibility to Cry1F, we predict the resistance phenotype of homozygous individuals for nonsense and frameshift mutations. This strategy provides higher sensitivity and should reduce labor and costs compared to traditional F1 and F2 screens to determine the frequency of diverse resistance alleles and, in specific mutations, susceptibility to Cry1F. Based on our observations, targeted sequencing overcomes the specificity limitations of Taqman genotyping, especially when targeting highly polymorphic genes. In addition, Taqman assays are limited to known alleles and cannot provide detection of novel candidate resistance alleles. Targeted sequencing provided preliminary evidence of a common mutation associated with Cry1F resistance in *S. frugiperda* from Brazil and Puerto Rico, which may suggest genetic flow between these populations as supported by whole genome-level comparisons [42]. Targeted sequencing such as Hi-Plex is adaptable to other pest-toxin resistance gene models, as long as known or suspected resistance loci are available, and the method can be expanded to sequence multiple genes within a single sample. While targeted sequencing provides an interesting list of candidate resistance alleles, further functional testing is needed to determine the role of missense mutations and their linkage with resistance to Cry1F.

Supplementary Materials: The following are available online at https://www.mdpi.com/article/10.3390/insects12070618/s1, Figure S1: Representative partial visualization of targeted sequencing reads aligned to the *SfABCC2* reference gene; Figure S2: Consensus SfABCC2 protein sequence used in this study; Table S1: List of sequenced *S. frugiperda* samples and their genotype for detected mutations; Table S2: List of TILING primers used in PCR amplification of *SfABCC2*.

Author Contributions: Conceptualization, K.S., K.L., S.J.E. and J.L.J.-F.; methodology, K.S., K.L., P.T., S.J.E., C.P.d.B., T.R. and J.L.J.-F.; software, K.S., K.L., P.T. and S.J.E.; formal analysis, K.S., K.L., S.J.E. and J.L.J.-F.; resources, D.M.V.D., A.M.L.-R. and J.L.J.-F.; data curation, K.S., K.L., S.J.E., C.P.d.B. and J.L.J.-F.; writing—original draft preparation, K.S., K.L., P.T., S.J.E. and J.L.J.-F.; writing—review and editing, all authors.; visualization, K.S., S.J.E. and J.L.J.-F.; funding acquisition, S.J.E. and J.L.J.-F. All authors have read and agreed to the published version of the manuscript.

Funding: This research was supported by Agriculture and Food Research Initiative Foundational Program competitive grant No. 2018-67013-27820 from the USDA National Institute of Food and Agriculture.

Institutional Review Board Statement: Not applicable.

Data Availability Statement: All targeted sequencing data used for this study are available through NCBI BioProject ID PRJNA729034.

Conflicts of Interest: The authors declare no conflict of interest.

References

1. Montezano, D.G.; Specht, A.; Sosa-Gómez, D.R.; Roque-Specht, V.F.; Sousa-Silva, J.C.; Paula-Moraes, S.V.; Peterson, J.A.; Hunt, T.E. Host plants of *Spodoptera frugiperda* (Lepidoptera: Noctuidae) in the Americas. *Afr. Entomol.* **2018**, *26*, 286–300. [CrossRef]
2. Westbrook, J.; Fleischer, S.; Jairam, S.; Meagher, R.; Nagoshi, R. Multigenerational migration of fall armyworm, a pest insect. *Ecosphere* **2019**, *10*, e02919. [CrossRef]
3. Nagoshi, R.N.; Fleischer, S.; Meagher, R.L.; Hay-Roe, M.; Khan, A.; Murua, M.G.; Silvie, P.; Vergara, C.; Westbrook, J. Fall armyworm migration across the Lesser Antilles and the potential for genetic exchanges between North and South American populations. *PLoS ONE* **2017**, *12*, e0171743. [CrossRef]
4. Storer, N.P.; Babcock, J.M.; Schlenz, M.; Meade, T.; Thompson, G.D.; Bing, J.W.; Huckaba, R.M. Discovery and characterization of field resistance to Bt maize: *Spodoptera frugiperda* (Lepidoptera: Noctuidae) in Puerto Rico. *J. Econ. Entomol.* **2010**, *103*, 1031–1038. [CrossRef] [PubMed]
5. Huang, F.; Qureshi, J.A.; Meagher, R.L., Jr.; Reisig, D.D.; Head, G.P.; Andow, D.A.; Ni, X.; Kerns, D.; Buntin, G.D.; Niu, Y.; et al. Cry1F resistance in fall armyworm *Spodoptera frugiperda*: Single gene versus pyramided Bt maize. *PLoS ONE* **2014**, *9*, e112958. [CrossRef]
6. Farias, J.R.; Andow, D.A.; Horikoshi, R.J.; Sorgatto, R.J.; Fresia, P.; dos Santos, A.C.; Omoto, C. Field-evolved resistance to Cry1F maize by *Spodoptera frugiperda* (Lepidoptera: Noctuidae) in Brazil. *Crop Protect.* **2014**, *64*, 150–158. [CrossRef]
7. Chandrasena, D.I.; Signorini, A.M.; Abratti, G.; Storer, N.P.; Olaciregui, M.L.; Alves, A.P.; Pilcher, C.D. Characterization of field-evolved resistance to *Bacillus thuringiensis*-derived Cry1F delta-endotoxin in *Spodoptera frugiperda* populations from Argentina. *Pest Manag. Sci.* **2018**, *74*, 746–754. [CrossRef] [PubMed]
8. Banerjee, R.; Hasler, J.; Meagher, R.; Nagoshi, R.; Hietala, L.; Huang, F.; Narva, K.; Jurat-Fuentes, J.L. Mechanism and DNA-based detection of field-evolved resistance to transgenic Bt corn in fall armyworm (*Spodoptera frugiperda*). *Sci. Rep.* **2017**, *7*, 10877. [CrossRef]
9. Flagel, L.; Lee, Y.W.; Wanjugi, H.; Swarup, S.; Brown, A.; Wang, J.; Kraft, E.; Greenplate, J.; Simmons, J.; Adams, N.; et al. Mutational disruption of the ABCC2 gene in fall armyworm, *Spodoptera frugiperda*, confers resistance to the Cry1Fa and Cry1A.105 insecticidal proteins. *Sci. Rep.* **2018**, *8*, 7255. [CrossRef] [PubMed]
10. Boaventura, D.; Ulrich, J.; Lueke, B.; Bolzan, A.; Okuma, D.; Gutbrod, O.; Geibel, S.; Zeng, Q.; Dourado, P.M.; Martinelli, S.; et al. Molecular characterization of Cry1F resistance in fall armyworm, *Spodoptera frugiperda* from Brazil. *Insect Biochem. Mol. Biol.* **2020**, *116*, 103280. [CrossRef]
11. Liu, Y.; Jin, M.; Wang, L.; Wang, H.; Xia, Z.; Yang, Y.; Bravo, A.; Soberón, M.; Xiao, Y.; Liu, K. SfABCC2 transporter extracellular loops 2 and 4 are responsible for the Cry1Fa insecticidal specificity against *Spodoptera frugiperda*. *Insect Biochem. Mol. Biol.* **2021**, *135*, 103608. [CrossRef]
12. Jin, M.-H.; Tao, J.-H.; Li, Q.; Cheng, Y.; Sun, X.-X.; Wu, K.-M.; Xiao, Y.-T. Genome editing of the *SfABCC2* gene confers resistance to Cry1F toxin from *Bacillus thuringiensis* in *Spodoptera frugiperda*. *J. Integr. Agric.* **2020**, in press. [CrossRef]
13. Abdelgaffar, H.; Perera, O.P.; Jurat-Fuentes, J.L. ABC transporter mutations in Cry1F-resistant fall armyworm (*Spodoptera frugiperda*) do not result in altered susceptibility to selected small molecule pesticides. *Pest Manag. Sci.* **2021**, *77*, 949–955. [CrossRef] [PubMed]
14. Vélez, A.M.; Spencer, T.A.; Alves, A.P.; Moellenbeck, D.; Meagher, R.L.; Chirakkal, H.; Siegfried, B.D. Inheritance of Cry1F resistance, cross-resistance and frequency of resistant alleles in *Spodoptera frugiperda* (Lepidoptera: Noctuidae). *Bull. Entomol. Res.* **2013**, *103*, 700–713. [CrossRef] [PubMed]
15. Jakka, S.R.; Knight, V.R.; Jurat-Fuentes, J.L. Fitness costs associated with field-evolved resistance to Bt maize in *Spodoptera frugiperda* (Lepidoptera: Noctuidae). *J. Econ. Entomol.* **2014**, *107*, 342–351. [CrossRef] [PubMed]
16. Nguyen-Dumont, T.; Teo, Z.L.; Pope, B.J.; Hammet, F.; Mahmoodi, M.; Tsimiklis, H.; Sabbaghian, N.; Tischkowitz, M.; Foulkes, W.D.; Giles, G.G.; et al. Hi-Plex for high-throughput mutation screening: Application to the breast cancer susceptibility gene PALB2. *BMC Med. Genom.* **2013**, *6*, 48. [CrossRef] [PubMed]
17. Nguyen-Dumont, T.; Pope, B.J.; Hammet, F.; Southey, M.C.; Park, D.J. A high-plex PCR approach for massively parallel sequencing. *BioTechniques* **2013**, *55*, 69–74. [CrossRef]
18. Storer, N.P.; Kubiszak, M.E.; Ed King, J.; Thompson, G.D.; Santos, A.C. Status of resistance to Bt maize in *Spodoptera frugiperda*: Lessons from Puerto Rico. *J. Invertebr. Pathol.* **2012**, *110*, 294–300. [CrossRef]

9.   Meagher, R.L. Trapping fall armyworm (Lepidoptera: Noctuidae) adults in traps baited with pheromone and a synthetic floral volatile compound. *Fla. Entomol.* **2001**, *84*, 288–292. [CrossRef]

20.   Capinera, J.L.; University of Florida. Fall Armyworm. Publication Number EENY-98. 2017. Available online: http://entnemdept. ufl.edu/creatures/field/fall_armyworm.htm#host (accessed on 20 May 2021).

21.   Blanco, C.A.; Portilla, M.; Jurat-Fuentes, J.L.; Sanchez, J.F.; Viteri, D.; Vega-Aquino, P.; Teran-Vargas, A.P.; Azuara-Dominguez, A.; Lopez, J.D.J.; Arias, R.; et al. Susceptibility of isofamilies of *Spodoptera frugiperda* (Lepidoptera: Noctuidae) to Cry1Ac and Cry1Fa proteins of *Bacillus thuringiensis. Southwest. Entomol.* **2010**, *35*, 409–415. [CrossRef]

22.   Gervais, A.L.; Marques, M.; Gaudreau, L. PCRTiler: Automated design of tiled and specific PCR primer pairs. *Nucleic Acids Res.* **2010**, *38*, W308–W312. [CrossRef] [PubMed]

23.   Zhang, J.; Kobert, K.; Flouri, T.; Stamatakis, A. PEAR: A fast and accurate Illumina Paired-End reAd mergeR. *Bioinformatics* **2013**, *30*, 614–620. [CrossRef] [PubMed]

24.   Bushnell, B. BBMap (BBMap Short Read Aligner, and Other Bioinformatic Tools). Available online: https://sourceforge.net/ projects/bbmap/ (accessed on 20 June 2019).

25.   Andrews, S. FastQC: A Quality Control Tool for High Throughput Sequencing Data [Online]. Available online: https://www. bioinformatics.babraham.ac.uk/projects/fastqc/ (accessed on 21 December 2017).

26.   Gimenez, S.; Abdelgaffar, H.; Goff, G.L.; Hilliou, F.; Blanco, C.A.; Hänniger, S.; Bretaudeau, A.; Legeai, F.; Nègre, N.; Jurat-Fuentes, J.L.; et al. Adaptation by copy number variation increases insecticide resistance in the fall armyworm. *Commun. Biol.* **2020**, *3*, 664. [CrossRef]

27.   Li, H. Aligning sequence reads, clone sequences and assembly contigs with BWA-MEM. *arXiv* **2013**, arXiv:1303.3997.

28.   Li, H. A statistical framework for SNP calling, mutation discovery, association mapping and population genetical parameter estimation from sequencing data. *Bioinformatics* **2011**, *27*, 2987–2993. [CrossRef]

29.   McKenna, A.; Hanna, M.; Banks, E.; Sivachenko, A.; Cibulskis, K.; Kernytsky, A.; Garimella, K.; Altshuler, D.; Gabriel, S.; Daly, M.; et al. The Genome Analysis Toolkit: A MapReduce framework for analyzing next-generation DNA sequencing data. *Genome Res.* **2010**, *20*, 1297–1303. [CrossRef]

30.   Slater, G.S.; Birney, E. Automated generation of heuristics for biological sequence comparison. *BMC Bioinform.* **2005**, *6*, 31. [CrossRef]

31.   Jombart, T.; Ahmed, I. Adegenet 1.3-1: New tools for the analysis of genome-wide SNP data. *Bioinformatics* **2011**, *27*, 370–3071. [CrossRef]

32.   Cingolani, P.; Platts, A.; Wang, L.L.; Coon, M.; Nguyen, T.; Wang, L.; Land, S.J.; Lu, X.; Ruden, D.M. A program for annotating and predicting the effects of single nucleotide polymorphisms, SnpEff: SNPs in the genome of *Drosophila melanogaster* strain w1118; iso-2; iso-3. *Fly (Austin)* **2012**, *6*, 80–92. [CrossRef]

33.   Chang, C.C.; Chow, C.C.; Tellier, L.C.; Vattikuti, S.; Purcell, S.M.; Lee, J.J. Second-generation PLINK: Rising to the challenge of larger and richer datasets. *Gigascience* **2015**, *4*, 7. [CrossRef] [PubMed]

34.   Wickham, H. *ggplot2: Elegant Graphics for Data Analysis*; Springer: New York, NY, USA, 2016.

35.   Delaneau, O.; Marchini, J.; Zagury, J.F. A linear complexity phasing method for thousands of genomes. *Nat. Methods* **2011**, *9*, 179–181. [CrossRef]

36.   Cheng, T.; Wu, J.; Wu, Y.; Chilukuri, R.V.; Huang, L.; Yamamoto, K.; Feng, L.; Li, W.; Chen, Z.; Guo, H.; et al. Genomic adaptation to polyphagy and insecticides in a major East Asian noctuid pest. *Nat. Ecol. Evol.* **2017**, *1*, 1747–1756. [CrossRef]

37.   Raj, A.; Stephens, M.; Pritchard, J.K. fastSTRUCTURE: Variational inference of population structure in large SNP data sets. *Genetics* **2014**, *197*, 573–589. [CrossRef] [PubMed]

38.   Niu, Y.; Qureshi, J.A.; Ni, X.; Head, G.P.; Price, P.A.; Meagher, R.L.; Kerns, D.; Levy, R.; Yang, X.; Huang, F. F2 screen for resistance to *Bacillus thuringiensis* Cry2Ab2-maize in field populations of *Spodoptera frugiperda* (Lepidoptera: Noctuidae) from the southern United States. *J. Invertebr. Pathol.* **2016**, *138*, 66–72. [CrossRef]

39.   Farias, J.R.; Andow, D.A.; Horikoshi, R.J.; Bernardi, D.; Ribeiro, R.D.S.; do Nascimento, A.R.; Dos Santos, A.C.; Omoto, C. Frequency of Cry1F resistance alleles in *Spodoptera frugiperda* (Lepidoptera: Noctuidae) in Brazil. *Pest Manag. Sci.* **2016**, *72*, 2295–2302. [CrossRef] [PubMed]

40.   Yainna, S.; Nègre, N.; Silvie, P.J.; Brévault, T.; Tay, W.T.; Gordon, K.; dAlençon, E.; Walsh, T.; Nam, K. Geographic monitoring of insecticide resistance mutations in native and invasive populations of the fall armyworm. *Insects* **2021**, *12*, 468. [CrossRef] [PubMed]

41.   Viteri, D.M.; Linares, A.M.; Cabrera, I.; Sarmiento, L. Presence of corn earworm and fall armyworm (Lepidoptera: Noctuidae) populations in sweet corn and their susceptibility to insecticides in Puerto Rico. *Fla. Entomol.* **2019**, *102*, 451–454. [CrossRef]

42.   Schlum, K.A.; Lamour, K.; de Bortoli, C.P.; Banerjee, R.; Meagher, R.; Pereira, E.; Murua, M.G.; Sword, G.A.; Tessnow, A.E.; Viteri Dillon, D.; et al. Whole genome comparisons reveal panmixia among fall armyworm (*Spodoptera frugiperda*) from diverse locations. *BMC Genom.* **2021**, *22*, 179. [CrossRef] [PubMed]

43.   Gutierrez-Moreno, R.; Mota-Sanchez, D.; Blanco, C.A.; Chandrasena, D.; Difonzo, C.; Conner, J.; Head, G.; Berman, K.; Wise, J. Susceptibility of fall armyworms (*Spodoptera frugiperda* j.e.) from mexico and puerto rico to Bt proteins. *Insects* **2020**, *11*, 831. [CrossRef] [PubMed]

44.   Vélez, A.M.; Spencer, T.A.; Alves, A.P.; Crespo, A.L.B.; Siegfried, B.D. Fitness costs of Cry1F resistance in fall armyworm, *Spodoptera frugiperda. J. Appl. Entomol.* **2013**, *138*, 315–325. [CrossRef]

Review

Resistance in the Genus *Spodoptera*: Key Insect Detoxification Genes

Frédérique Hilliou [1] , **Thomas Chertemps** [2], **Martine Maïbèche** [2] and **Gaëlle Le Goff** [1,*]

[1] Université Côte D'Azur, INRAE, CNRS, ISA, F-06903 Sophia Antipolis, France; frederique.hilliou@inrae.fr
[2] Institut D'Ecologie et des Sciences de L'Environnement de Paris, Sorbonne Université, CNRS, INRAE, IRD, iEES-Paris, F-75005 Paris, France; thomas.chertemps@sorbonne-universite.fr (T.C.); martine.maibeche@sorbonne-universite.fr (M.M.)
* Correspondence: gaelle.le-goff@inrae.fr; Tel.: +33-492-386-578

Simple Summary: The moth larvae are among the most damaging pest species on crops worldwide. In this review, we focus on the genus *Spodoptera*, which can feed on many crops such as rice, cotton or corn. The massive use of insecticides to control these insects has led to the development of resistance. Here, we aim to compare the resistance mechanisms of four species (*Spodoptera exigua*, *Spodoptera frugiperda*, *Spodoptera littoralis* and *Spodoptera litura*) and highlight the role of enzymes and transporters in resistance to help us understand the molecular basis of their origin.

Abstract: The genus *Spodoptera* (Lepidoptera: Noctuidae) includes species that are among the most important crop pests in the world. These polyphagous species are able to feed on many plants, including corn, rice and cotton. In addition to their ability to adapt to toxic compounds produced by plants, they have developed resistance to the chemical insecticides used for their control. One of the main mechanisms developed by insects to become resistant involves detoxification enzymes. In this review, we illustrate some examples of the role of major families of detoxification enzymes such as cytochromes P450, carboxyl/cholinesterases, glutathione S-transferases (GST) and transporters such as ATP-binding cassette (ABC) transporters in insecticide resistance. We compare available data for four species, *Spodoptera exigua*, *S. frugiperda*, *S. littoralis* and *S. litura*. Molecular mechanisms underlying the involvement of these genes in resistance will be described, including the duplication of the CYP9A cluster, over-expression of GST epsilon or point mutations in acetylcholinesterase and ABCC2. This review is not intended to be exhaustive but to highlight the key roles of certain genes.

Keywords: resistance; *Spodoptera*; cytochromes P450; carboxyl/cholinesterases; glutathione S-transferases; ATP-binding cassette transporters

Citation: Hilliou, F.; Chertemps, T.; Maïbèche, M.; Le Goff, G. Resistance in the Genus *Spodoptera*: Key Insect Detoxification Genes. *Insects* **2021**, *12*, 544. https://doi.org/10.3390/insects12060544

Academic Editor: Béla Darvas

Received: 18 May 2021
Accepted: 8 June 2021
Published: 11 June 2021

1. Introduction

The genus *Spodoptera* (Lepidotera: Noctuidae) contains some of the most important insect crop pests, many of which are highly polyphagous species, able to feed on more than 100 host plants including maize, rice, cotton (e.g., *Spodoptera litura* (Fabricius) [1]). They are present on all continents and their potential invasiveness has been highlighted in recent years, notably with the species *S. frugiperda* (J.E. Smith). Native to the American continent, it was detected in Africa in 2016 [2] and has since invaded Asia and Australia (CABI, Wallingford, UK, 2021). With its flight capabilities [3,4] and under favorable climatic conditions, its invasion of Europe in the near future seems inevitable.

The control of these pests requires the massive use of insecticides. They have developed resistance to all chemical families and three of the four *Spodoptera* species present in the Arthropod Pesticide Resistance Database are in the top 15 most resistant arthropods on the planet: *S. litura*, *S. frugiperda* and *S. exigua* (Hübner) [5].

Table 1 shows the available data on the molecules for which these three species, as well as *S. littoralis* (Boisduval) have developed resistance. These insects have developed resistance to all the chemical families: organophosphates, carbamates, pyrethroids but also for a more recent family such as diamides. There are usually two main mechanisms in insecticide resistance, either target modification or mechanisms that reduce the amount of insecticide reaching the target (reduced penetration, sequestration or intervention of detoxification enzymes). Detoxification involves enzymes that catalyze successive reactions to make the insecticidal molecule less toxic and more easily excreted from the body. Cytochromes P450 (P450) and carboxyl/cholinesterases (CCE) are phase I enzymes that catalyze oxidation, hydrolysis and reduction. Glutathione S-transferases (GST) are phase II enzymes and catalyze the addition of a group such as glutathione. This step is called conjugation and is followed by phase III excretion involving ATP-binding cassette (ABC) transporters. These proteins use the energy of ATP hydrolysis to transport substrates across lipid membranes. The resistance caused by the enzymes involved in these three phases implies either a modification of their level of expression or their catalytic activity. The molecular process behind this can be of several kinds. A point mutation in the sequence of a gene can modify the catalytic activity of the enzyme. For example, in the mosquito *Anopheles funestus*, the increased activity of GSTe2 is due to a point mutation (L119F) that leads to an increase in the accessibility of the active site, allowing high resistance to DDT [6]. Overexpression is one of the other commonly demonstrated mechanisms, involving duplications, amplifications or cis-or trans-regulations [7]. A well-known example of gene duplication is the case of *CYP6G1* in *Drosophila melanogaster*. The overexpression of this gene in multi-insecticide resistant populations is due to duplication as well as insertion of transposable elements in its promoter [8]. In the aphid *Myzus persicae*, it is a gene amplification of one or two F4 and FE4 esterases that is at the origin of the resistance to organosphosphates, carbamates and pyrethroids [9]. Up to 80 copies of the same gene are found in some resistant aphids. The esterase in this case represented up to 3% of the total proteins [10]. A recent example illustrates cis and trans regulation in a chlorpyrifos-resistant strain of *S. exigua* [11]. The authors showed that the resistance was mainly due to the overexpression of a P450, CYP321A8, a P450 capable of metabolizing chlorpyrifos, cypermethrin and deltamethrin. This overexpression is due to both constitutive overexpression of the transcription factors Cap'n'collar isoform C (CncC) and Maf (Muscle aponeurosis fibromatosis, trans-regulation), major regulators of detoxification in insects [12], and to a mutation in the promoter of this P450. The mutation creates a cis-regulatory element that promotes binding of a Knirps nuclear receptor.

This review highlights some examples that demonstrate the involvement of each of the four families of detoxification genes (P450, CCE, GST and ABC) in insecticide resistance and the molecular mechanisms identified for pests of the genus *Spodoptera*. We have chosen to focus on specific genes that we consider emblematic rather than an exhaustive review of their roles.

Table 1. Assessment of insecticide resistance worldwide for the four species of *Spodoptera*, data extracted from the Arthropod Pesticide Resistance Database (http://www.pesticideresistance.org, accessed on 18 May 2021).

Insecticide Chemical Class	*S. exigua*	*S. frugiperda*	*S. littoralis*	*S. litura*
Avermectins	Abamectin (#31), emamectin benzoate (#48)			Abamectin (#43), emamectin benzoate (#27)
Benzoylurea	Chlorfluazuron (#9), diflubenzuron, lufenuron (#10), teflubenzuron	Lufenuron (#2), triflumuron	Diflubenzuron, teflubenzuron (#3)	Lufenuron (#8)
Bacillus thuringensis	Bt var. unspecified (#10), var aizawai, var. kurstaki		Bt var unspecified (#3), var aizawai	
Bt toxins	Cry1Ca	Cry1Aa, Cry1A.105, Cry1Ab (#2), Cry1Ac (#3), Cry1F (#54), Cry2Ab2 (#2), Vip3A (#3)		

Table 1. *Cont.*

Insecticide Chemical Class	*S. exigua*	*S. frugiperda*	*S. littoralis*	*S. litura*
Carbamates	Methomyl (#18), thiodicarb	Carbaryl (#7), methomyl (#6), thiodicarb (#2)	Carbaryl (#2), methomyl (#2),	Carbaryl (#2), methomyl (#38), thiodicarb (#33)
Cyclodienes	BHC/cyclodiene—unspecified in literature (#3), endosulfan (#19)	Aldrin, dieldrin, lindane (#2)	Endrin, toxaphene (#2)	Endosulfan (#31), lindane
Diamides	Chlorantraniliprole (#26), cyantraniliprole, flubendiamide (#5)	Chlorantraniliprole (#2), flubendiamide (#2)		Chlorantraniliprole (#11)
Diacylhydrazines	Methoxyfenozide (#25), tebufenozide (#19)		Tebufenozide	Methoxyfenozide (#36), tebufenozide
Neonicotinoids				Acetamiprid
Organochlorine	DDT (#4)	DDT (#3)	DDT	DDT (#2)
Organophosphates	Chlorpyrifos (#48), parathion-methyl (#3), profenofos (#22), quinalphos (#9)	Acephate, chlorpurifos (#7), diazinon (#2), dichlorvos, malathion (#2), parathion-methyl (#4), sulprofos, trichlorfon,	Acephate, azinphos-methyl, chlorpyrifos (#4), fenitrothion, leptophos, methamidophos, methidathion, monocrotophos (#4), parathion (#2), parathion-methyl (#3), profenofos, sulprofos, triazophos, trichlorfon	Chlorfenvinphos, chlorpyrifos (#55), diazinon, dichlorvos, malathion, monocrotophos (#5), phoxim (#4), profenofos (#56), quinalphos (#9), triazophos (#2), trichlorfon
Oxadiazines	Indoxacarb (#44), metaflumizone (#4)		Indoxacarb	Indoxacarb (#50), metaflumizone
Phenylpyrazoles				Fipronil (#15)
Pyrethroids	Bifenthrine (#13), cyfluthrin (#2), cyhalothrin-lambda (#2), cypermethrin (#56), cypermethrin-beta, deltamethrin (#43), fenpropathrin (#14), fenvalerate (#5), permethrin (#2), pyrethroids-unspecified in literature (#3)	Bifenthrine, cyfluthrin, cyhalothrin, cyhalothrin-lambda (#11), cypermethrin (#2), cypermethrin-zeta, deltamethrin (#2), fenvalerate, fluvalinate, permethrin (#5), tau-fluvalinate, tralomethrin	Cypermethrin (#4), deltamethrin (#2), fenvalerate, flucythrinate, permethrin	Bifenthrine (#33), cyfluthrin (#12), cyfluthrin-beta (#9), cyhalothrin (#3), cyhalothrin-lambda (#11), cypermethrin (#49), cypermethrin-beta (#15), deltamethrin (#50), esfenvalerate (#9), fenpropathrin (#10), fenvalerate (#7), pyrethrins
Pyrroles	Chlorfenapyr (#11)			Chlorfenapyr (#5)
Spinosyns	Spinetoram, spinosad (#56)		Spinetoram (#2), spinosad (#2)	Spinosad (#39)

number of cases reported in Arthropod Pesticide Resistance database.

2. Phase I: Functionalization

Once a xenobiotic enters the cell, it is processed by enzymes that make it functional. The phase I enzymes carry out oxidation reactions, hydrolysis, etc., making the metabolite more polar and more easily excreted.

2.1. Cytochrome P450s

Cytochrome P450 genes (CYP) constitute one of the largest gene families with representatives in nearly all living organisms [13]. CYPs catalyze a large number of reactions including hydroxylation, epoxidation, oxidation [14]. In insects, CYPs are involved in endogenous metabolism such as the biosynthesis of ecdysteroid [15] or the production of hydrocarbon [16], as well as in detoxification mechanisms with the metabolization of xenobiotics (e.g., plant secondary metabolites and insecticides). Their number varies from 36 in the louse *Pediculus humanus* [17] to 196 in the mosquito, *Culex quinquefasciatus* [18]. Nomenclature has been proposed based on sequence homology [19,20]. P450s belong to the same family designated by an Arabic numeral when they share 40% identity and to the same subfamily designated by capital letters when this percentage is higher than

55%. The nomenclature was revised and the notion of clan was proposed as a higher leve of classification to take into account the increasing number of available sequences [21 Initially, the arthropod P450s were divided into four clans, but a recent study brings thi number to 6 with the historical clans, clan 2, clan 3, clan 4, mitochondrial clan to whic clans 16 and 20 were added [22]. Clans 3 and 4 are the two main clans known to hav P450s involved in resistance mechanisms, but not exclusively, for example, *CYP12D1* fron *D. melanogaster* which belongs to the mitochondrial clan is over-expressed in resistar strains [23] and has been shown to confer resistance to DDT and dicyclanil [24]. We wi highlight the role of some of P450s in insecticide resistance and evolution of the *CYP9.* subfamily within the four *Spodoptera* species.

2.1.1. Phylogeny of *CYP9A*

CYP9A belongs to clan 3. In Lepidoptera, *CYP9As* were identified as a cluster in comparative study of BAC sequences from three species *Bombyx mori*, *Helicoverpa armiger* and *S. frugiperda* [25] with the presence of four, five and nine genes respectively, showin the rapid evolution of the cluster. In the two more closely related species (*H. armigera* an *S. frugiperda* which diverged about ±20 MYA), only three orthologous pairs were ident fied [26]. The availability of the genome of the most closely related species in the genu *Spodoptera* enabled the evolution of this group to be assessed. In *S. frugiperda*, obtainin the chromosomally resolved genome of the maize strain helped identify 14 *CYP9A* gene (genome Version 6 [27]), 12 of which were in a cluster (Figure 1), i.e., three more than th nine genes previously identified in the BAC study, whereas 15 genes were identified i the rice strain (genome version 3 [28]). These numbers are very similar to those foun in other species, 15 genes in *S. litura* (Figure 1) [1], 12 genes in *S. exigua* and 11 gene in *S. littoralis* (genome in progress, transcriptome available). In the latter, the number i potentially underestimated and improved genome resolution could lead to increases.

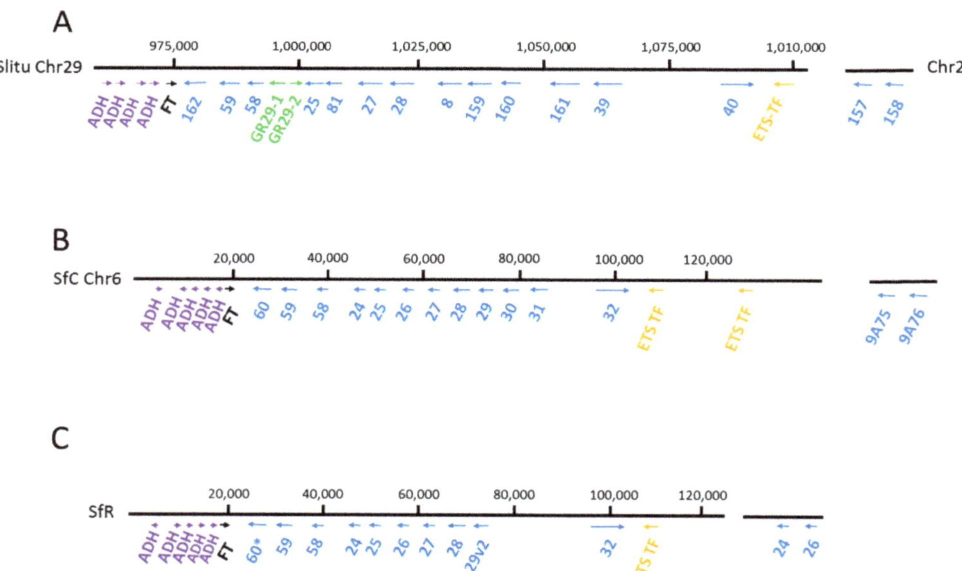

Figure 1. CYP9A cluster in the genome of *S. litura* (**A**), *S. frugiperda* corn strain (**B**) and *S. frugiperda* rice strain (**C**). Numbers in blue correspond to *CYP9A* genes. Arrows correspond to gene orientation, ADH: alcohol dehydrogenase, FT: fucosyl transferase, EST TF: E26 transformation specific transcription factor, * pseudogene.

Among these genes, eight were true orthologs, *CYP9A25*, *CYP9A26*, *CYP9A27*, *CYP9A2 CYP9A30*, *CYP9A31*, *CYP9A58* and *CYP9A59* in *S. frugiperda*, *S. litura* and *S. littoralis*, prob ably already clustered in their common ancestor (Figure 2). *CYP9A24* is only found in *S. frugiperda*. Some genes of the cluster are *S. frugiperda* corn strain specific (*CYP9A32*, *CYP9A60*, and *CYP9A76*), *S. frugiperda* rice strain specific (*CYP9A91*) or *S. litura* specific (*CYP9A39-157-158-161-162*). Orthologs of *CYP9A26*, *CYP9A59* and *CYP9A27* were also found in the genome of *S. exigua*.

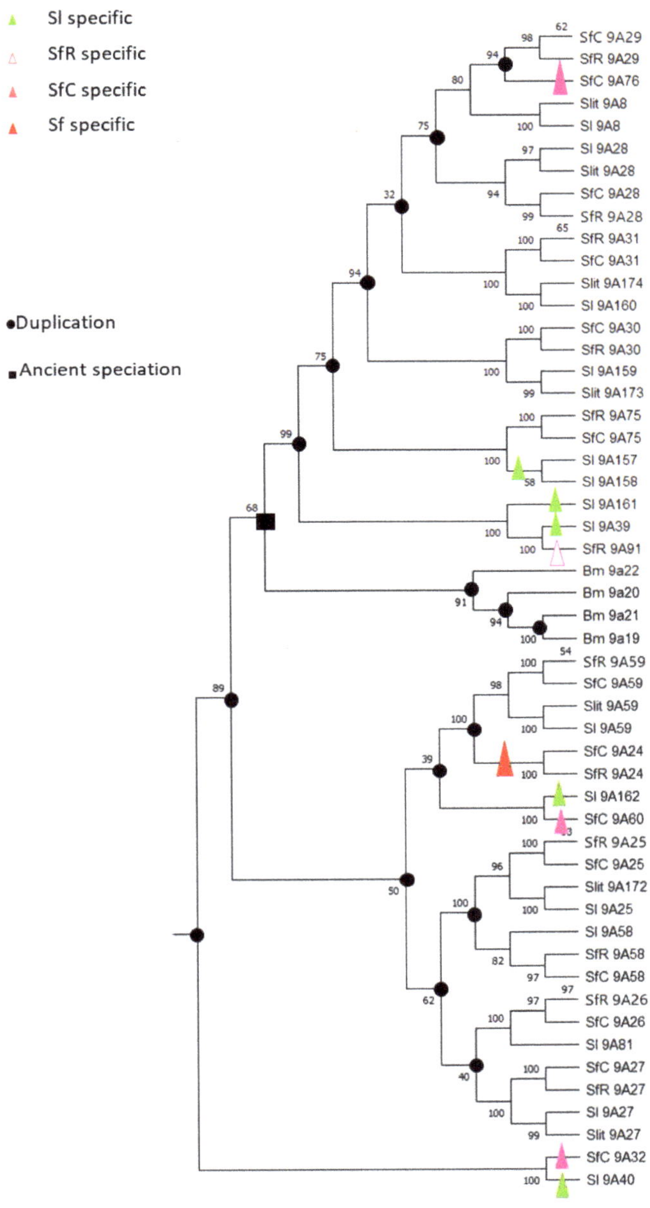

Figure 2. *CYP9A* subtree: ML phylogeny with bootstrap value. SfC: *S. frugiperda* corn strain, SfR: *S. frugiperda* rice strain, Sl: *S. litura*, Slit: *S. littoralis*, Bm: *B. mori*. Slit9A58, Slit9A81, and Slit9A40-like are missing from the tree. Black square: ancient speciation representing *B. mori* split from *Spodoptera* species. Black dot: duplication events.

The synteny of the CYP9 cluster is conserved between *S. frugiperda* (corn and rice variants) and *S. litura* (Figure 1) where the same genes, alcohol dehydrogenase (*ADH*), fucosyl transferase (*FT*) and E26 transformation specific (*ETS*, a transcription factor) were found on each side of the cluster. In *S. litura*, two gustatory receptors from the GR29 family are part of the *CYP9* cluster. They form a head to tail tandem between *CYP9A58* and *CYP9A25* and they replace *CYP9A24*. The GR29 family is specific to *Spodoptera* and based on the evolutionary history of *S. litura* and *S. littoralis* [29], we expect GR genes to be present in the *CYP9* cluster of *S. littoralis*. The mechanisms resulting in blooming in the *CYP9A* family in the *Spodoptera* complex still remain to be deciphered. Several origins are possible, one of which corresponds to transposable elements as they have been shown to be prevalent in the surrounding of CYP genes in Lepidoptera [25,30,31], *Drosophila* [32] or mosquito [33]. Other mechanisms include duplications (tandem, chromosome or genome duplications) and retropositions [26]. This *CYP9A* bloom likely provides *Spodoptera* species with a selective advantage and potentially diverse catalytic capabilities for each of these enzymes. Indeed, induction experiments show that within the *CYP9A* cluster, genes have their own induction profile depending on the xenobiotics used. For example, *CYP9A31* is the most induced gene in response to xanthotoxin treatment in both *S. frugiperda* and *S. litura* whereas *CYP9A28* is not induced by this molecule but by indole [34,35]. This suggests that within the cluster, each gene has its own catalytic competence unrelated to its position in the cluster or phylogeny as it has been suggested for the *CYP6AE* cluster in *H. armigera* [36]. Dermauw et al., (2020) suggested that there would be a selective advantage to keeping the cluster as a heritable unit, which would enable adaptation to new environments [22].

2.1.2. Resistance through Over-Expression of *CYP9A*

We next examine CYP9As in the light of insecticide resistance: the first *CYP9A* was found in *Heliothis virescens* and named *CYP9A1* [37]. It was in a thiodicarb-resistant population of *H. virescens*. An elevated level of *CYP9A1* mRNA was detected in the resistant strain compared to the susceptible strain. Several subsequent studies have associated over-expression of certain *CYP9As* with insecticide resistance, but very few studies have gone so far as to demonstrate their involvement in insecticide metabolism. An example came from *H. armigera*, where pyrethroid resistance was associated with constitutive over-expression of P450s in the laboratory-selected YGF strain [38]. In this strain, *CYP9A12* and *CYP9A14* were over-expressed 433-and 59-fold, respectively, in the fat body. The functional expression of these two P450s in *Saccharomyces cerevisiae* demonstrated their capacity to metabolize a pyrethroid insecticide, esfenvalerate [39]. Elevated levels of *CYP9As* associated with resistance have also been shown in *Spodoptera* spp. In a recent study, a Brazilian population of *S. frugiperda* resistant to *Bacillus thuringiensis* (Bt) Cry1 toxins was tested against 15 insecticide molecules with different modes of action. In addition to resistance to Bt toxins, the Sf-Des strain had showed 14- and eight-fold resistance to deltamethrin and chlorpyrifos, respectively [40]. P450 activity was increased in the Sf-Des strain compared to the susceptible Bt strain. RNAseq experiments confirmed the over-expression of several P450s, including *CYP9As* and RT-qPCR analysis validated that a *CYP9A-like* was expressed more than 200-fold in the resistant strain. In another study on a laboratory-selected population of *S. frugiperda* with lufenuron, a *CYP9A-like* gene was over-expressed 45-fold [41]. In both of these studies, the cause of the over-expression of the *CYP9As* was not identified, whereas Gimenez et al., (2020) reported the existence of two copies of the full CYP9A cluster of in a Puerto Rican resistant strain (PR) of *S. frugiperda*. PR was resistant to deltamethrin and the use of a P450 synergist, piperonil butoxide (PBO) abolished the resistance, confirming the primary role of P450. The *CYP9A* cluster locus is under positive selection in this strain [27]. Several field-collected strains of *S. exigua* resistant to organophosphates, pyrethroids and diamides also exhibited an over-expression of *CYP9A*. For a chlorpyrifos-resistant strain, the resistance was mainly associated with *CYP321A8*, which is over-expressed and capable of metabolising insecticides, however monitoring the expression of 68 P450s by RT-qPCR showed significant over-expression of several *CYP9As* including *CYP9A11*, *CYP9A27*, *CYP9A97* and *CYP9A98* [11]. Are these genes also involved in resistance? This remains to be demonstrated. However, there is some evidence of their involvement in resistance, such as the RNAi knockout (KO) of *CYP9A98*, which makes *S. exigua* larvae more sensitive to deltamethrin [42], the knockout of *CYP9A10*, which makes them more sensitive to alpha-cypermethrin [43], and the knockout of *CYP9A105*, which increases mortality to deltamethrin, alpha-cypermethrin, and fenvalerate [44]. Similar results were also obtained in the species *S. litura*. Indeed, strains resistant to fenvalerate, beta-cypermethrin and cyhalothrin had their resistance reduced by the use of PBO [45]. RNAseq experiments for these field-collected Chinese strains (LF and NJ) showed over-expression of several P450s including *CYP9As*. *CYP9A40* was over-expressed 663.4- and 76.13-fold in LF and NJ, respectively, compared with the susceptible strain. Previous experiments showed that dsRNA silencing of *CYP9A40* increased the sensitivity of *S. litura* to deltamethrin [46]. To our knowledge, no data suggesting a link between insecticide resistance and *CYP9A* of *S. littoralis* are available at this time. There is sufficient evidence across closely related species to suggest a major role for *CYP9As* at least in pyrethroid resistance; however, the ability of these enzymes to metabolise insecticides remains to be demonstrated. The cause of the over-expression of these genes also remains to be identified. In one case, it is due to the duplication of the *CYP9A* cluster [27] while in the study of Hu et al., (2021) it can be assumed that the over-expression of the transcription factor CncC, already shown in several insects to be the major regulator of detoxification genes, leads to the over-expression of *CYP9A*.

2.2. Carboxyl/Cholinesterases (CCEs)

Insect carboxylesterases or carboxyl/cholinesterases (CCEs) have a wide range of physiological functions across living organisms, from metabolism of endogenous compounds (hormones, pheromones, neurotransmitters) to detoxification of various xenobiotics. They play a critical role in defense against various allelochemicals associated with plants and insecticides [47]. In particular, CCEs have been implicated in resistance to pyrethroids (PYRs), organophosphates and carbamates (CBs) in numerous pest species [48,49]. The mechanisms of esterase-mediated resistance involve either metabolic resistance or target site mutation [49]. Metabolic resistance can be based either on insecticide sequestration or insecticide hydrolysis: overproduction of CCEs through gene amplification or transcriptional up-regulation can lead to insecticide molecule sequestration without any (or very slow) hydrolysis of the insecticide, whereas point mutations can alter the catalytic properties of some CCEs, leading to increased hydrolysis towards insecticides. Target site resistance is due to mutation of acetylcholinesterases (AChEs) that renders them less sensitive to inhibition by insecticides.

Insect CCEs fall into three main functional groups, representing dietary/detoxification, hormone/semiochemical processing, and neuro/developmental functions [50]. A comprehensive phylogenetic analysis of CCE sequences isolated from *H. armigera*, *H. zea*, *B. mori* and *Manduca sexta* genomes has recently refined the CCE nomenclature in Lepidoptera [51]. The neuro/developmental class comprises seven clades (027 to 033) with generally catalytically incompetent proteins, excepted for AChEs (clade 027). The hormone/semiochemical processing class (clades 020 to 026) includes among others, juvenile hormone esterases, several pheromone-processing esterases, but also some CCEs (clade 026) previously associated with organophosphate (OP) resistance in Hemiptera (namely E4, FE4, [10]). Finally, the dietary/detoxification group (clades 001 to 019 and 034) is the most diversified: some enzymes have been implicated in semiochemical processing but most of them have been associated with dietary and detoxification functions, including insecticide resistance for some dipteran and hymenopteran CCEs (the α-esterases, reviewed in [47]). CCEs from this third group are the most abundant in Lepidoptera and in particular the three most numerous clades (001, 006 and 016) with species-specific radiations. Many clades in this third group are also Lepidoptera specific [51]. The total number of CCE genes identified so far in *S. litura* and *S. frugiperda* genomes were 110 and 96, respectively [1,28]. For *S. exigua*, we were able to retrieve 73 sequences from public database (GenBank:GC. 011316535.1). For *S. littoralis*, we have previously identified 30 CCE transcripts expressed in antennae [52–54]. This repertoire was supplemented with 26 new sequences by Walker et al., (2019) [55]. Searching in our RNAseq databases allowed us to identify an additional 19 CCE transcripts (Table S1), leading to a total of 65 SlitCCE sequences, which is likely underestimated compared to its sister species *S. litura*.

Although comprehensive complete repertoires of CCEs are now available in many lepidopteran species, including spodopterans, still very few CCE genes have been directly linked with insecticide resistance. Here, we will focus on some spodopteran CCEs for which a direct role in resistance is supported either by in vitro and/or in vivo approaches i.e., enzymes both identified at the molecular level and directly involved in insecticide hydrolysis or insecticide susceptibility.

2.2.1. Resistance through Over-Expression of CCE

As reviewed previously in Farnsworth et al., (2010), there is a strong correlation between insecticide resistance and higher esterase activities in the four *Spodoptera* species studied here [56]. Esterase activities toward artificial substrates in vitro (generally α-or β-naphthyl acetate) were compared between homogenates of susceptible and resistant strains, which allowed for the detection of both higher overall CCE activities or higher staining intensities of certain isozymes after analysis by native PAGE electrophoresis. For example, higher staining intensities of several esterases isozymes have been associated with OP and PYR resistance in *S. littoralis* [57], *S. litura* [58], *S. exigua* [59] and *S. frugiperda* [60]. The underlying biochemical mechanisms are not yet known, but it is hypothesized that increased sequestration of insecticides by CCEs and catabolism apply [56].

In lepidopterans, over-expression of CCE associated with insecticide resistance has been intensively studied in *H. armigera* [61–63], especially through proteomic and real-time polymerase chain reaction (RT-PCR) approaches. Most of the over-expressed CCE identified at the molecular level belonged to the 001 clade, which is very diverse in this species, with 21 CCE001 genes. This clade also shows a large expansion in spodopterans, with 23 and 19 CCE001 genes annotated in *S. litura* [1] and *S. frugiperda* [28], respectively. We counted 14 CCE001 sequences in *S. littoralis* ([52–55] Table S1), a number which is probably again underestimated in the absence of genome annotation. In *S. litura*, some of these CCE001 genes were inducible by imidacloprid [1]. Most interestingly, knock-down of two SlituCEE001s (SlituCOE57 and SlituCOE58) by siRNA injection increased the sensitivity of *S. litura* larvae to imidacloprid when fed with an insecticide-containing diet [1]. This is the first demonstration of a direct link between CCE induction and insecticide sensitivity at the molecular level in a lepidopteran species. After BLAST (Basic Local Alignment Search Tool) searches, we found orthologous sequences of SlituCOE57 and SlituCOE58 in *S. littoralis* (SlitCXE50 and SlitCXE48, respectively, Table S1) and *S. frugiperda* (SfruCCE001i and SfruCCE001f, respectively, [55]), suggesting that they might play a similar role in insecticide resistance. In *S. litura*, all of these CCE001s were grouped into a large cluster on chromosome 2 [1], as previously observed for the CCE001 of *H. armigera* [51].

2.2.2. Metabolic Resistance through Point Mutations of CCEs

Structural mutations in the active site of a carboxylesterase could result in a reduction in the ability of this enzyme to hydrolyze common carboxylesterase substrates (such as naphthyl acetate esters) but convert it to an OP hydrolase. This corresponds to the mutant ali-esterase mechanism [64]. Metabolic resistance by point mutation in CCE sequence has been observed for quite some time in several non-lepidopteran species, mostly in Diptera [49,65,66]. Two point mutations, G137D and W251L (or G151D and W271L as reported in Cui et al., (2011) following the sequence of *D. melanogaster* AChE [67]), were indeed found in field-resistant populations of several species, including *Lucilia cuprina* and *Musca domestica*, and involved in altering the substrate specificity of CCE, leading to increased activity towards OPs [65–67]. In addition to dipterans, another amino acid change at position 251 (W251G) was also found in an OP-resistant strain of the parasitoid wasp *Anisopteromalus calandrae* [68].

To the best of our knowledge, no point mutation-based resistance in CCE has been resolved at the molecular level in resistant Lepidopteran strains. However, an in vitro study conducted on CCEs from four insect orders including Lepidoptera, tested whether the change of substrate specificity associated with these two mutations could be a more general feature in OP-resistant insects [67]. The catalytic properties of seven CCEs mostly belonging to the dietary/detoxification group were analyzed, including one CCE from *S. litura* (GenBankEU783914). Recombinant mutant proteins were tested in vitro towards two OPs (paraoxon and chlorfenvinphos) and β-naphthyl acetate. For the seven enzymes tested, the G151D and W271L mutations conferred OP activity in 62.5% and 87.5% cases, respectively [67]. The same in vitro approach was developed more recently on eight *H. armigera* CCEs but with more contrasting results [62], with increased insecticide hydrolysis being observed only for some enzymes. However, for one esterase (HarmCCE001c), PYRs (fenvalerate and cypermethrine), hydrolysis was enhanced 14-fold after the leucine mutation. The *S. litura* CCE tested by [67] matched the SlituCOE082 genomic sequence identified later [1]. We found sequences orthologous to SlituCOE082 in *S. frugiperda* and *S. littoralis*, but not in *S. exigua*. SlituCOE82 presented 95.7% of amino acid identity with SlitCXE64 (Table S1) and 91.4% with SfruCCE16j [55], with residues conserved at the putative mutation positions (Figure 3).

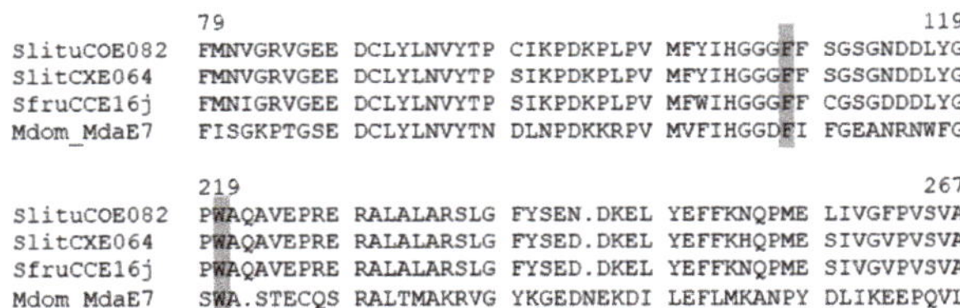

Figure 3. Alignment of two 50 amino acid sequence portions of SlituCOE082, SlitCXE64, SfruCCE16j and MdaE7 from *Musca domestica* (Rutgers diazinon-R resistant strain). The putative mutation positions G151D and W271L are shaded in grey. According to [67], the amino acid corresponding to G137D and W251L were G107 and W219 in *S. litura* sequence; for *S. litura* carboxyl/cholinesterase (CCE), the mutation tested in vitro was G151D. Genbank accession number, *S. frugiperda* (XP_035450257.1), *S. litura* (XP_022828113.1), *M. domestica* (AAD29685.1).

SlituCOE082 belongs to the clade 016 which is also expanded in Lepidoptera [51]. It includes 16 genes in *S. litura*, 14 in *S. frugiperda*, eight in *S. exigua* and 16 in *S. littoralis* (including nine additional sequences from RNAseq data). Phylogenetic analysis of these 54 sequences (Figure 4) illustrates their diversification within the four Spodoptera species, and in particular the 1:1 orthologous relationship is clear for seven sequences from each species. In *S. litura*, the CCEs from in clade 016 showed a major expansion on chromosome 25, with a cluster of 12 adjacent sequences [1]. Six genes in this cluster showed induction by neonicotinoid insecticide, i.e., imidacloprid in *S. litura* larvae. In particular, SlituCOE82 is moderately induced by imidacloprid in Malpighian tubules [1].

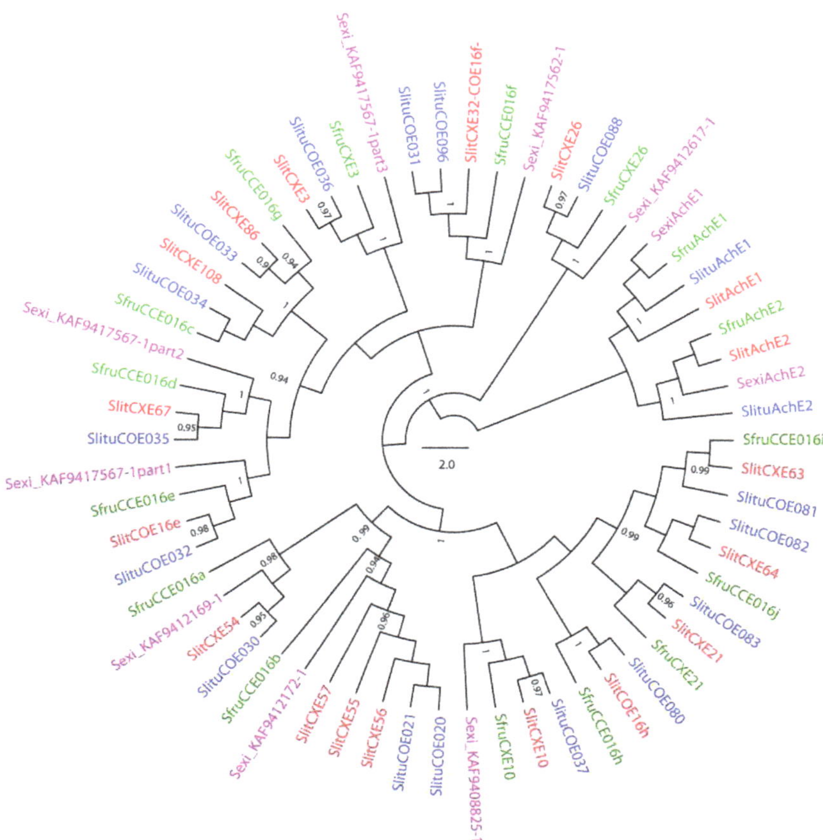

Figure 4. Maximum-likelihood phylogeny of spodopteran CCEs from Clade (016) and acetylcholinesterases (AChEs). The tree was built from amino-acid sequences of CCE repertoires of *S. littoralis* (branches colored in red), *S. frugiperda* Corn strains (green), *S. litura* (blue), *S. exigua* (purple). Sequences from [1,28] for *S. litura* and *S. frugiperda*, respectively. For *S. littoralis*, from [52,55] and new sequence from our RNAseq database (Table S1). For *S. exigua*, GenBank numbers were inserted the sequence name, SexiAChE1 (AZB49078.1) and SexiAChE2 (AZB49079.1). Likelihood-ratio test values are only indicated in the internal nodes defining orthogroups, when aLRT > 0.9. The scale bar represents 0.2 expected amino-acid substitutions per site.

To date, the presence of the G151D and/or W271L mutations in field-resistant spodopteran strains has not been found. However, although metabolic resistance by overproduction of CCE could induce broader resistance than qualitative mutation, the two studies discussed below suggest that point mutations could be a more common mechanism for insecticide resistance than expected. A systematic comparison of the corresponding CCE gene sequences in susceptible and resistant lepidopteran pest strains would be necessary to assess their precise role in resistance.

2.2.3. Target Site Resistance through Point Mutations; the Case of Acetylcholinesterases

Acetylcholinesterase (AChE or ace) terminates nerve impulses by catalyzing the hydrolysis of acetylcholine in cholinergic synapses. Irreversible inhibition of AChE by CBs and OPs thus causes acetylcholine to accumulate in synapses and acetylcholine receptors to open permanently, resulting in insect death. Most insects have two *AChE* genes but only the gene expressed in the central nervous system (namely ace-1 or *AChE1*) is essential for synapse functioning [69]. All four *Spodoptera* species possess two *AChE* genes as expected

(Figure 4) with a clear orthologous relationship and highly conserved sequences (more than 98% of amino acid identity). Target site resistance mediated by *AChE1* insensitivity to insecticides has been identified by biochemical approaches in several insect species, and subsequently elucidated at the molecular level for some species [49,69]. Sequencing of *AChE* genes from field-resistant strains and comparison with susceptible strains described several point mutations conserved across species, most of which modify the active site of the enzyme. Three substitutions (A201S, G227A and F290V, numbering corresponding to the mature enzyme of *Torpedo californica*) were first reported in CB- and/or OP-resistant strains of aphids (*A. gossypii* [70]), dipterans (*Bactrocera dorsalis* [71]) and lepidopterans (*Chilo suppressalis* [72] and in *P. xylostella* [73]).

The involvement of acetylcholinesterases in OP/CB resistance has been demonstrated in biochemical approaches on *S. litura* resistant strains from Korea [74] or India [75,76]. Similarly, in *S. exigua*, the AChE enzyme of a carbamate resistant strain from California was found to be approximately 30-fold more insensitive to methomyl compared to the enzyme from the susceptible laboratory strain [77]. However, within Spodoptera species, molecular data on the AChE point mutation are mostly available for *S. frugiperda*. In this species, Yu et al., (2003) showed that acetylcholinesterase of a field strain collected from corn fields in Florida was clearly far less sensitive (up to 85-fold) than that of a susceptible strain to inhibition by CBs and OPs [60]. Kinetic data also showed that the apparent Km value of acetylcholinesterase from the field strain was 56% of that from the susceptible strain. By comparing the predicted amino acid sequences of a *S. frugiperda* ace-1 fragment from a chlorpyrifos susceptible strain with the corresponding sequence isolated from an 18.1-fold resistant strain from Brazil, Carvalho et al., (2013) were then able to identify the three previously described point mutations (Figure 5) [78].

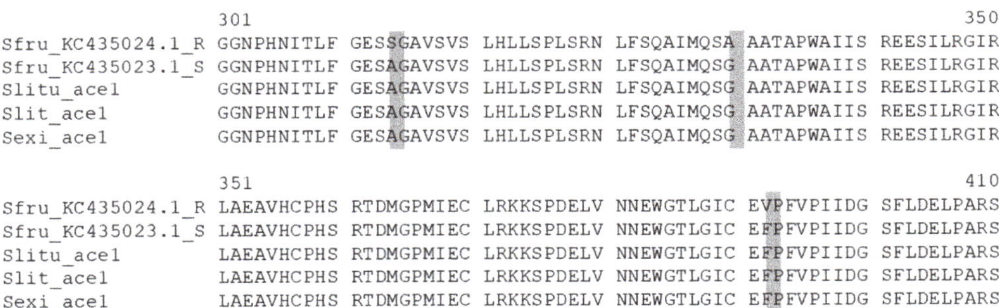

Figure 5. Alignment of a 120 amino acid sequence portion of AChE1 *S. frugiperda* susceptible (Genbank accession number KC435023.1) and resistant strain (KC435024.1) with orthologous sequences from *S. litura* (XP_022819835.1 = SlituCOE002), *S. littoralis* (Table S1) and *S. exigua* (AZB49078.1). The mutation positions are shaded in grey.

Genotyping revealed that of the three, the A201S allele was present at relatively low frequency (17.5%) while G227A and F290V were present at higher frequency (67.5% and 32.5%, respectively) [78]. More recently, Boaventura et al., (2020) searched for the presence of these mutations in 34 different resistant populations of *S. frugiperda* collected from four different continents [79] and showed that F290V was the most frequent substitution in all populations tested. Similar results were obtained by Guan et al., (2020) on *S. frugiperda* individuals collected from China and Africa, except that only two positions (A201S, F290V) were found, with the F290V allele at higher frequencies [80].

3. Phase II: Conjugation

Following oxido-reduction and hydrolysis reactions performed by phase I enzymes, xenobiotic metabolites (including insecticides) are then conjugated to small hydrophilic molecules by phase II enzymes, a group of transferases that metabolize hydrophobic compounds containing nucleophilic or electrophilic groups [81].

3.1. Glutathione-S-Transferases (GSTs)

Glutathione-S-transferases (GSTs, EC 2.5.1.18) are one of the most important classe of this group whose biotransformation reaction leads to the generation of hydrophil metabolites, which are readily excreted by efflux transporters such as ABC transporter In arthropods, GSTs are highly diverse and are involved in a variety of cellular function from the detoxification of a wide range of both endogenous and xenobiotic compounds, t intracellular transport, hormone biosynthesis and reduction of oxidative stress [82]. GS1 primarily catalyze the conjugation of electrophilic lipophilic compounds with the thio group of reduced glutathione (GSH) but are also capable of catalyzing a dehydrochlorina tion reaction using reduced glutathione as a cofactor. The enzymatic structure of cytosoli GSTs classically consists of hetero- or homo-dimeric proteins, with each monomer consis ing of a highly conserved amino-terminal domain providing the GSH-binding site (G-site while the carboxyl terminal domain interacts with the hydrophobic substrate (H-site) [82

In insects, cytosolic GSTs belong to a diverse gene family divided into six classe (Delta, Epsilon, Omega, Sigma, Theta, and Zeta) based on their substrate specificitie and phylogenetic relationships [83]. The growing number of available genomes reveals large disparity in the total number of GSTs among insect species, with diversity rangin from low in hymenopterans (eight GSTs in *Apis mellifera*) to high in dipterans (39 in *Cule quinquefasciatus*) [84]. This variability is mainly due to genetic expansions observed in th Insecta-specific Delta and Epsilon classes due to multiple duplication events [85].

Insect GSTs have been intensively studied for their role in insecticide resistanc reports correlating high levels of GST activity with high resistance to various pesticide including organophosphates, organochlorines, cyclodienes, and pyrethroids, exist for man species and among them *S. littoralis* and *S. frugiperda* [86]. This phenomenon relies o the precise regulation or induction of GST expression by xenobiotic compounds, and i particular the Epsilon class (GSTe) [87,88]. Here, we will focus on spodopteran GSTes tha have been shown to contribute to insecticide response and resistance.

3.1.1. Phylogeny of GST Epsilon

GST-mediated resistance could be triggered by different mechanisms, including gen amplification by multiple duplication events could lead to enhanced detoxification o insecticides underlying of the resistance process. Based on the recent availability o chromosome-level assembly genomes from spodopterans, we re-annotated GST epsilo class. For convenience, we reconciled the previous nomenclature with our new annotatio to generate a unified dataset (see Supplementary Materials Table S2) [1,28]. Overall, GST accounts for nearly half of the total number of GSTs identified in the Spodopteran genome (20 over 40 on average). Our phylogenetic analysis (Figure 6) revealed clear orthologou relationships among the four species studied with 1:1 orthologs in almost all cases, an with epsilon GSTs of *H. armigera*, suggesting that this great diversification is occurring i the Noctuoidea clade, and especially in highly polyphagous species.

Such diversity could be explained by tandem and segmental gene duplications a demonstrated by the phylogenetic branching that corresponds to the genomic clustering o GSTe (Figure 6). Furthermore, synteny studies support this diversification, with half of th GSTe derived from only two duplication events, arguing that these clades share a commo GST ancestor (in an ancient Lepidoptera ancestor) even though their genetic expansion occurred independently (Figure 7, [1]).

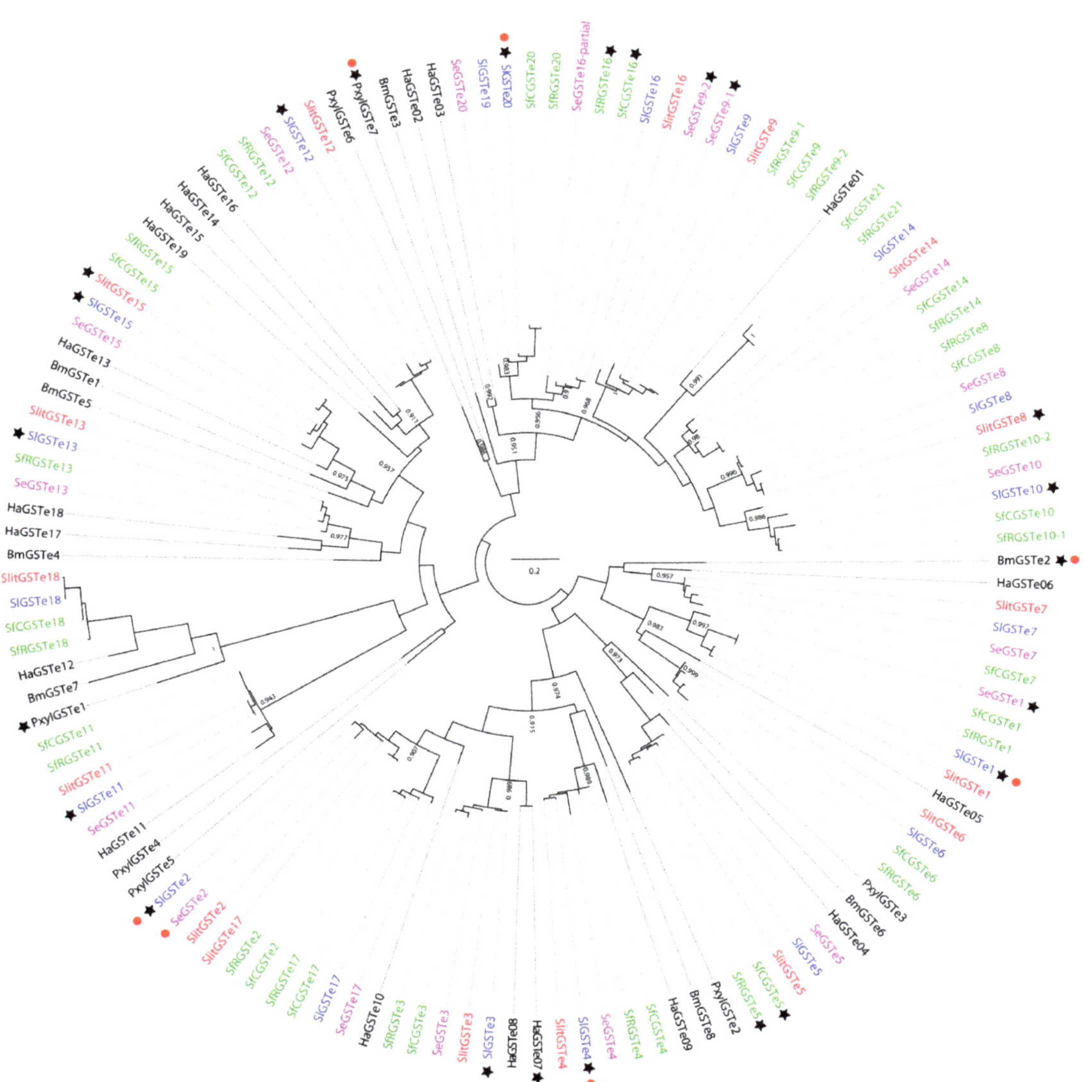

Figure 6. Maximum-likelihood phylogeny of lepidopteran epsilon glutathione-S-transferases (GSTes). The tree was built from amino-acid sequences of GST repertoires of *S. littoralis* (branches colored in red), *S. frugiperda* Corn and Rice strains (green), *S. litura* (blue), *S. exigua* (purple), *H. armigera*, *P. xylostella* and *B. mori* (black). Likelihood-ratio test values are only indicated in the internal nodes defining orthogroups, when aLRT > 0.9. Black stars indicate that GSTes are overexpressed in insecticide-resistant strains or when exposed to pesticides. Red dots indicate GSTes with demonstrated in vitro/in vivo insecticide resistance. The scale bar represents 0.2 expected amino-acid substitutions per site.

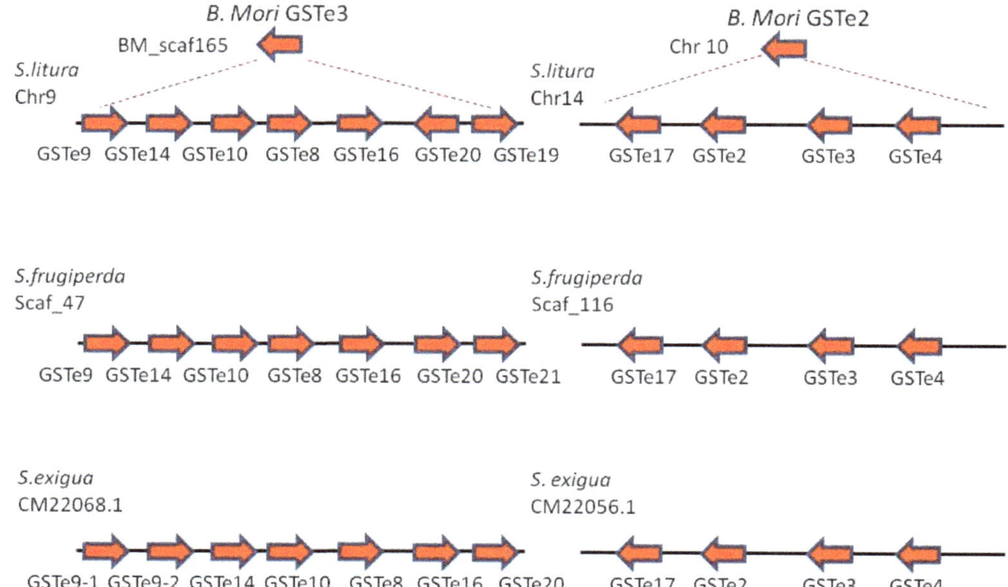

Figure 7. GSTe clusters derived from *B. mori* GSTe2 and GSTe3 identified in the genomes of *S. litura*, *S. frugiperda* corn and rice strains and *S. exigua* (adapted from [1]).

This extensive genetic amplification does not directly support a demonstration of insecticide resistance per se; however, this highly diverse class is expected to have ancient roles unique to polyphagous lepidopterans, such as the removal of harmful chemical compounds from natural sources. According to the pre-adaptation hypothesis, phytophagous insect species with more diverse diets are likely to acquire resistance to more diverse insecticides [89], and thus genetic diversity may underlie the origin and evolution of insecticide resistance.

3.1.2. GST Activity against Insecticides

The biochemical mechanisms of GST-based insecticide resistance are classically associated with the conjugation of GSH to the pre-inactivated phase I compound. Such conjugation resistance has been demonstrated for different classes of organophosphate or pyrethroid insecticides [48,82]. Moreover, dehydrochlorination reactions using GSH as a cofactor have been implicated in resistance to the organochlorine DDT [6] making GSTs potential phase I enzymes for hydrochloric compounds. In addition to these direct modes of action, GST has also been shown to reduce lipid hydroperoxides produced as a result of insecticide-induced damages [90]. This peroxidase activity of GSTs protects tissues from an excess of reactive oxygen species (ROS), which can be more harmful than the pesticide itself [91]. Finally, GSTs are also capable of simple binding and sequestration activity towards various compounds, such as pyrethroids [86].

In spodopterans, most in vitro insecticide studies rely on the use of competitive 1-chloro-2,4-dinitrobenzene (CDNB) assays, a method that compares the amount of CDNB, the universal substrate of GST, in the presence or absence of a competitor, i.e., an insecticide. While this technique clearly indicates an interaction between a GST and a pesticide, it does not decipher the mechanism by which this interaction occurs. *S. litura* SlGSTe2 and SlGSTe3 were the first spodotperan GSTes to be characterized in this manner, showing differential activity toward DDT and deltamethrin between the two enzymes, with SlG-STe2 having the greater activity toward the insecticide [92]. The SeGSTe2 ortholog was further characterized showing activity toward metaflumizone, indoxacarb, monosultap, chlorpyrifos, malathion, cyhalothrin, and imidacloprid, suggesting a conserved insecticide metabolizing function [93]. Similarly, GSTe1 was carefully analyzed in *S. litura* and *S. exigua* species [94,95]. CDNB competition assays showed activity towards chlopyrifos, malathion, phoxim, deltamethrin and cypermethrin and the specific activities of GSTE1 were further analyzed using high-performance liquid chromatography (HPLC), confirming their active binding to chlorpyrifos and cypermethrin. Interestingly, GSTe1 is also active towards secondary plant metabolites [96] and has strong peroxidase activity. Therefore, this enzyme has multi-detoxification properties with a wide range of substrates, potentially conferring insecticide resistance.

In vivo studies have also confirmed the direct involvement of GSTs in insecticide tolerance. In Cheng et al., (2019), injection of siRNAi against SlGSTe20 and SlGSTe3 increased the sensitivity of *S. litura* larvae to imidacloprid, and the recombinant proteins subsequently showed activity towards diazinon, permethrin, chlorfenapyr, and bendiocarb [97].

3.1.3. Resistance through Over-Expression of GST Epsilon

GST overexpression has often been a prerequisite for the identification of enzymes potentially involved in insecticide resistance [48]. Thus, many studies have identified candidate GSTes whose expression could be altered upon exposure to insecticides. In *S. litura*, SlGSTe1, e2, e3, e4, e10, e11, e12, e13, e15 and e20 showed overexpression following exposure to various pesticides such as tebufenozide, carbaryl, DDT, malathion, deltamethrin, chlorpyrifos and imidacloprid [1,92,98,99]. It is noteworthy that among these 10 over-expressed genes, no specific phylogeny-related pattern could be assigned (Figure 6) indicating that, rather than defining an insecticide-specific clade, the induced GSTe genes are diverse with potentially redundant and/or complementary activities. Similar patterns are observed in *S. littoralis* and *S. exigua* when exposed to deltamethrin or lambda-cyhalothrin, chlorpyrifos and chlorantraniliprole, respectively [94,100]. Nevertheless, induction after exposure may be only part of the animal's response to a given toxicant and does not necessarily imply involvement in resistance [90]. Therefore, in an effort to find genes directly associated with insecticide catabolism, studies focusing on the expression profiles of genes highly expressed in resistant strains compared to the susceptible population have indicated potential candidates in GSTe as directly responsible for insecticide resistance. In *S. frugiperda*, SfGSTe5 was associated with an OP-resistant strain using a transcriptomic approach [78]. SfGSTe5 is the ortholog of *Plutella xylostella* PxylGSTe3, an epsilon GST induced and capable of metabolizing the organophosphate insecticides parathion and methylparathion [101,102]. In an RNAseq experiment comparing indoxacarb-resistant and susceptible *S. litura* populations [103], SlGSTe4 and SlGSTe20 were the only GSTs overexpressed. Interestingly, both of these GSTes are active against imidacloprid, diazinon, permethrin, chlorfenapyr, or bendiocarb [97,101], indicating potential cross-activity of SlGSTe for various insecticides. This complex pattern of gene expression regulation must be orchestrated by specific regulatory operators that can effectively govern down-or overexpression upon exposure but could also be affected by mutations in resistant strains to account for specific activations. Promoter sequence analysis revealed that some *S. litura* GSTs harbor the same cap 'n' collar 'C/muscle aponeurosis fibromatosis (CncC/Maf) binding site and the aryl hydrocarbon receptor/nuclear aryl hydrocarbon receptor translocator (AhR/ARNT) binding site [84,94]. These transcription factors coordinately regulate GST expression through intracellular production of ROS induced by insecticide exposure. In a CncC RNAi experiment, Shi et al., (2020) demonstrate that this transcription factor not only enhances indoxacarb sensitivity in susceptible and resistant strains of *S. litura*, but is also involved in the down-regulation of detoxification genes related to indoxacarb resistance, confirming its central role in insecticide-resistance mechanisms [104].

3.1.4. Resistance through Point Mutations in GST Epsilon

Insecticide resistance by GSTe has been clearly established in *Anopheles* where a key residue, Phe120, has been implicated in the increased binding of DDT to GSTe2 [6]. Such a substitution is also observed in 5 GSTes from spodopteran species (GSTe2, e12, e15, e17, e1 Figure 8) indicating potential conservation of resistance processes between these speci€ Thus, further investigations are needed to fully understand the detailed mechanisms : pesticide-GSTes interaction and their precise contribution to insecticide resistance.

```
                110                      120                      130                      140
SlGSTe18    D  A  G  I  L  F  P  -  I  D  S  A  I  F  S  D  F  F  A  G  K  -  -  -  W  P  A  N  E  V  L  I  N  K
SlitGSTe18  D  A  G  I  L  F  P  -  I  D  S  A  I  F  S  D  F  F  A  G  K  -  -  -  W  P  A  N  E  V  L  I  N  K
SfCGSTe18   D  A  G  I  L  F  P  -  I  D  S  A  I  F  S  D  L  F  A  G  K  -  -  -  W  P  A  N  E  V  L  I  N  K
SfRGSTe18   D  A  G  I  L  F  P  -  I  D  S  A  I  F  S  D  L  F  A  G  K  -  -  -  W  P  A  N  E  V  L  I  N  K
AfunGSTe2   E  S  G  V  L  F  A  R  M  R  F  I  F  E  R  I  L  F  Y  G  K  S  D  I  P  E  D  R  V  E  Y  V  Q  K
SlGSTe15    D  S  G  I  L  Y  P  A  L  R  E  N  D  E  P  I  F  F  G  T  A  T  S  L  K  P  E  G  A  A  K  I  R  S
SlitGSTe15  D  S  G  I  L  Y  P  A  L  R  E  N  D  E  P  I  F  F  G  T  A  T  S  L  K  P  E  G  V  A  K  I  R  S
SfCGSTe15   D  S  G  I  L  Y  P  A  L  R  E  N  D  E  P  I  F  F  G  T  A  T  S  L  K  P  E  G  E  A  K  I  R  S
SfRGSTe15   D  S  G  I  L  Y  P  A  L  R  E  N  D  E  P  I  F  F  G  T  A  T  S  L  K  P  E  G  V  A  K  I  R  S
SeGSTe15    D  S  G  I  L  Y  P  A  L  R  E  N  D  E  P  I  F  F  G  T  A  T  S  L  K  P  G  G  V  A  K  I  K  S
SlGSTe12    D  S  G  I  L  F  P  S  L  R  G  T  V  E  P  V  L  F  W  G  E  K  S  F  R  Q  E  N  L  D  K  I  K  K
SlitGSTe12  D  S  G  I  L  F  P  S  L  R  G  T  V  E  P  V  L  F  W  G  E  K  S  F  R  Q  E  N  L  D  K  I  K  K
SfCGSTe12   D  S  G  I  L  F  P  S  L  R  G  T  V  E  P  V  F  F  L  G  E  K  S  F  R  E  E  N  L  A  K  I  K  K
SfRGSTe12   D  S  G  I  L  F  P  S  L  R  G  T  V  E  P  V  F  F  L  G  E  K  S  F  R  E  E  N  L  A  K  I  K  K
SeGSTe12    D  S  G  I  L  F  P  S  L  R  G  T  V  E  P  V  L  F  W  G  E  T  S  F  R  P  E  N  L  A  K  I  K  K
SlGSTe2     N  V  G  I  F  F  I  R  L  K  V  V  V  L  P  A  I  F  G  D  L  D  G  P  T  E  Q  H  K  A  D  I  D  E
SlitGSTe2   N  V  G  I  F  F  I  R  L  K  V  V  V  L  P  A  I  F  G  D  L  D  G  P  T  E  K  H  K  A  D  I  D  E
SlitGSTe17  N  V  G  I  F  F  I  R  L  K  V  V  V  L  P  A  I  F  G  D  L  D  G  P  T  E  K  H  K  A  D  I  D  E
SfCGSTe2    N  V  G  I  F  F  I  R  L  K  V  I  V  L  P  A  I  F  G  D  L  D  G  P  T  E  K  H  K  A  D  I  D  E
SfRGSTe2    N  V  G  I  F  F  I  R  L  K  V  I  V  L  P  A  I  F  G  D  L  D  G  P  T  E  K  H  K  A  D  I  D  E
SeGSTe2     N  V  G  I  F  F  I  R  L  K  V  I  V  L  P  A  L  F  G  D  L  D  G  P  T  E  K  H  K  A  D  I  D  E
SlGSTe17    N  A  G  V  F  F  L  K  L  L  G  C  V  L  P  A  V  F  G  D  L  D  G  P  T  Q  Q  H  K  A  E  I  D  A
SeGSTe17    N  A  G  V  F  F  L  K  L  L  G  C  I  L  P  A  V  F  G  D  L  D  G  P  T  E  Q  H  K  L  E  I  D  A
SfCGSTe17   N  A  G  V  F  F  L  K  L  L  G  C  V  L  P  A  V  F  G  D  L  D  G  P  T  E  Q  H  K  A  E  I  D  A
SfRGSTe17   N  A  G  V  F  F  L  K  L  L  G  C  V  L  P  A  V  F  G  D  L  D  G  P  T  E  Q  H  K  A  E  I  D  A
```

Figure 8. Alignment of a subset of GSTes from *S. litura* (Sl) *S. frugiperda* corn and rice strains (SfC and SfR), *S. exigua* (Se), *S. littoralis* (Slit) and AfunGSTe2 from *Anopheles funestus* (Afun, genbank accession number KC800363.1) harboring a L119F substitution. The L119F mutation (in red) is associated with DDT resistance in Anophelines [6].

4. Phase III: Elimination/Export

The last phase concerns the elimination or excretion of metabolites outside the ce Metabolites have been rendered less toxic by the reactions they underwent in phase and/or phase II but they can also be directly excreted without prior chemical modificatio This phase mainly involves ATP-binding cassette transporters.

4.1. ATP-Binding Cassette Transporters (ABCs)

ATP-binding cassette (ABC) transporters are proteins that use the energy of A⁻ hydrolysis to transport substrates such as amino acids, lipids, peptides, sugars and dru€ across cell membranes. ABC transporters are classified into eight subfamilies from let€ A to H based on similarities in their ATP binding domain. Full transporters (FT) cons s of two cytosolic nucleotide-binding domains (NBDs) and two transmembrane domai- (TMDs) whereas half transporters (HT) have only one of each domain and must dimeri: to form a functional transporter. The NBD is involved in ATP binding and hydrolysis. T- TMD consists of 5-6 transmembrane segments and is responsible for substrate specificity. I insects, their number varies from 32 in *Nilaparvata lugens* [105] to 82 in *Plutella xylostella* [1C6 while in the available genomes of Noctuidae about 50 ABCs have been found so far. So- subfamilies are conserved well from humans to arthropods. For example, ABCD, ABC. ABCF have clear orthology and similar suspected physiological roles such as the transpo of acyl-CoA molecules into the peroxisome, ribosome biogenesis and translation. Moreove some ABC transporters subfamilies are involved in chemotherapy resistance in humans an€ pesticides resistance in insects. Their names reflect these roles, the B subfamily is know as multidrug-resistance protein (MDR) or P-glycoprotein (P-gp) while the C subfamily i known as multidrug-resistance associated protein (MRP). For example, the human MD⁊ (ABCB1) excludes chemotherapeutic agents and confers resistance in several types o cancer when over-expressed [107]. MDR1 orthologs in *Drosophila Mdr50* and *Mdr65* ar over-expressed in DDT-resistant strain 91-R. Knockdown of each gene in this strain ℩ RNAi increases the susceptibility of flies to DDT [108]. These examples, in humans an€ *Drosophila*, illustrate one of the mechanisms of resistance involving ABC transporters: ov=

expression of the transporter which allows exclusion of a greater amount of xenobiotics. An additional mechanism in the case of insecticide resistance can be mentioned and corresponds to the mutation of the target. Indeed, ABCs are one of the receptor proteins of *Bacillus thuringiensis* (Bt) toxins as are cadherin-like, aminopeptidase-N or alkaline phosphatase, and a modification in the sequence of these receptors can confer resistance (for a recent review see [109]).

4.1.1. Resistance through Point Mutations in ABCs, the Case of ABCC2

ABCC2 was first identified as a receptor for one of the Bt crystal toxins (Cry), Cry1Ac in the lepidopteran *Heliothis virescens* [110]. In this study, a strain of *H. virescens* was selected in the laboratory for several years with Cry1Ab and Cry1Ac toxins. Very high levels of resistance were obtained as well as a loss of membrane binding for the toxins. A deletion in exon 2 of *ABCC2* resulting in a 99 amino acid protein truncation was identified as the cause of resistance to Cry1Ac [110]. This study paved the way to investigating the role of ABC transporters in resistance to Bt-toxin. Special attention is given to the four species of interest in this review. Annotations of ABC transporters in the genome of *S. litura, S. frugiperda, S. exigua* and in the transcriptomic data of *S. littoralis* were mainly performed automatically and only a few subfamilies were annotated manually by experts (G. Le Goff, personal communication). In a recent publication, Liu et al., (2018) showed that the sensitivity towards Cry1Ac toxin differed by a factor of 65 between the more tolerant *S. litura*, and *S. frugiperda* [111]. SlABCC2 and SfABCC2 share 97% identity. Fragment substitutions and point mutations in the transporter expressed in insect cells allowed the authors to identify the amino acid at position 125 as the key to this difference in sensitivity. A glutamine is present at this position in *S. frugiperda* while it is a glutamic acid in *S. litura*. By aligning the sequences of ABCC2 between the four *Spodoptera* species (*S. exigua, S. frugiperda, S. littoralis* and *S. litura*), it is possible to predict their susceptibility to Cry1Ac toxin. *S. exigua* should be susceptible like *S. frugiperda* while *S. littoralis* should be more tolerant like *S. litura* (Figure 9).

S. exigua ABCC2	110 RTVQPLLFSQLLSYWSVDSD 129
S. frugiperda ABCC2	110 RTAQPLLFSQLLSYWSVDSE 129
S. littoralis ABCC2	110 RTVQPLLFSELLSYWSVDSE 129
S. litura ABCC2	110 RTVQPLLFSELLSYWSVDSE 129

Figure 9. Alignment of a 20 amino acid sequence of ABCC2 from four *Spodoptera* species, presence of a glutamine or glutamic acid at position 119. Genbank accession number, *S. exigua* AIB06822, *S. frugiperda* AUO38740, *S. littoralis*, *S. litura* XM_022967434.

Three-dimensional structure predictions suggest that the G or E at position 125 is localized in the extracellular loop 1 (ECL1). Liu et al., hypothesize that it may have a role in toxin binding, which would explain these differences in sensitivity [111]. The importance of ECL1 in Bt toxin selectivity was previously demonstrated in *B. mori*, where replacement of certain amino acids in the ECL1 loop between BmABCC2 and BmABCC3 resulted in increased binding affinity of Cry1A toxins [112]. These variations in toxin receptor sequence may also explain the spectrum of specificity of Bt toxins. Indeed, some toxins primarily target lepidopterans, for example Cry1 toxins, while others are specific to Coleoptera, such as Cry3 [113].

Larger changes in the sequence of ABCC2 (Figure 10A) have been associated with resistance in some of these *Spodoptera* spp., especially in field-resistant *S. frugiperda* populations. Indeed, resistance to transgenic maize expressing the Cry1Fa toxin has been reported since 2007 in Puerto Rico for *S. frugiperda*. Banerjee and colleagues have demonstrated that the resistance was due to mutations in the SfABCC2 toxin receptor [114]. These mutations result in a truncated protein that loses the second transmembrane domain and at the same time toxin binding. A nine-base deletion (position 39–47) and a two-base insertion (GC at position 2218) lead to a frameshift and the occurrence of a premature stop codon. The truncated protein is 746 amino acids (Figure 10B) whereas the ABCC2 of the susceptible strain encodes a protein of 1349 amino acids (Figure 10A). The frequency of mutated *SfABCC2* in Puerto Rico increased between 2007 and 2009 from 1% to 42% and stabilized at 55% in 2017. This truncated receptor was not found by the authors in the populations from Florida or the Dominican Republic [114].

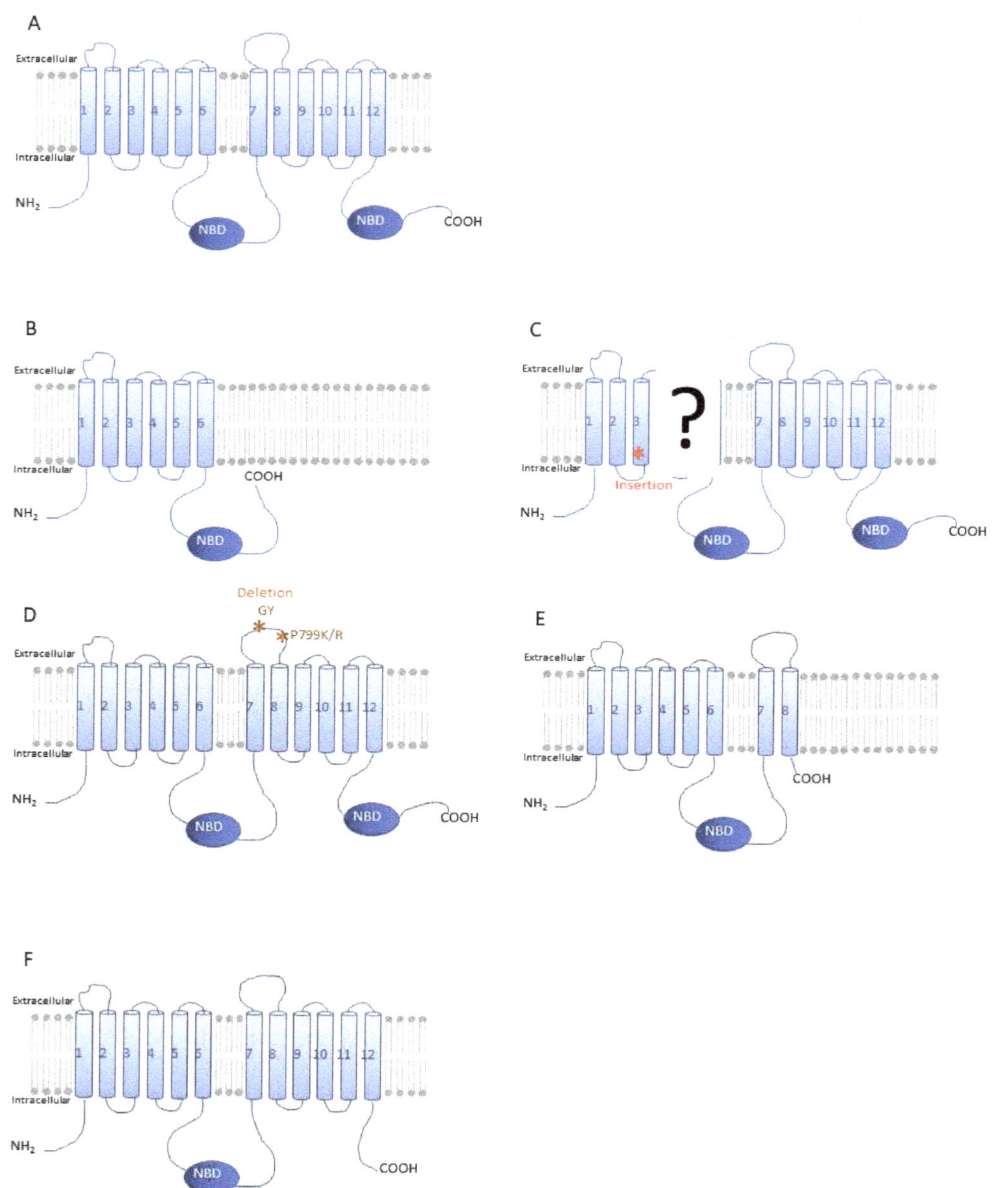

Figure 10. Protein structure of ABCC2 of *Spodoptera* spp from susceptible or resistant populations. (**A**) Protein structure of SfABCC2 and SeABCC2 wild-type. (**B,C**) Protein structures of SfABCC2 in resistant populations from Puerto Rico, * corresponds to an insertion of amino acids [114,115]. (**D,E**) Protein structure of SfABCC2 in resistant populations from Brazil [80,116]. (**F**) Protein structure of SeABCC2 from a laboratory selected strain [117].

Another group confirmed these results. Flagel et al., isolated a population of *S. frugiperda* from a site in Puerto Rico in 2010 and this population was selected in the laboratory with Cry1Fa for 50 generations [115]. The resulting strain had a resistance factor of 500 for Cry1Fa and 87-fold cross-resistance to Cry1A.105 compared to the susceptible reference strain. They found the same GC insertion in the ABCC2 sequence that creates a stop codon at position 747 and results in a truncated protein. By conducting a sampling campaign in various localities of Puerto Rico and Brazil, the mutation was to be found in Puerto Rico only. In addition to this allele, the authors mention the existence of a second resistance allele and the insertion of a sequence in the fourth exon of the gene. However, sequencing difficulties due to repeated sequences did not allow them to identify precisely the effect of this insertion on the *ABCC2* sequence (Figure 10C) although an aberrant splicing of this allele 2 was suspected.

Furthermore, a recent study reported resistance to Cry1F toxin in isolated Brazilian populations of *S. frugiperda* [116]. The authors identified two mutations in the extracellular loop 4 (ECL4) of ABCC2: a deletion of two amino acids (GY) at positions 788 and 789 and the change of a proline to either lysine or arginine at position 799 (Figure 10D). Unlike the resistance reported in Puerto Rico, these mutations do not cause premature termination of the sequence. Expression of mutated ABCC2 in insect cells confirmed the role of these mutations in toxin binding. Analysis of populations collected in different regions of Brazil showed a high frequency of the GY deletion as well as many rare alleles that result in sequence changes between amino acids 783–799. According to the authors, these results reinforce the primary role of the ECL4 extracellular loop in Cry1F toxicity. Further analysis is needed to demonstrate the role that these mutations may have in resistance to other toxins. Indeed, in addition to resistance to Cry1F other studies on Puerto Rican populations with truncated SfABCC2 have shown cross-resistance for Cry1A.105, Cry1Ac, and Cry1Ab toxins [114,115]. Do mutations in the ECL4 loop also confer these cross-resistances? Another mutation in ABCC2 has been identified in Brazilian populations corresponding to the insertion of 12 bases at the intracellular loop between transmembrane domains 8 and 9 resulting in a premature stop codon [80]. The truncated protein lost the last four transmembrane segments and the second intracellular ATP-binding domain (Figure 10E). The study was performed on insects collected three years after those of Boaventura et al., (2020) and changes in the use of transgenic crops may have occurred. The authors did not investigate the effect of this mutation on resistance and of Bt toxin binding. Instead, they tested whether invasive populations of *S. frugiperda* (also sampled in Africa and Asia) carried mutations known to cause insecticide resistance.

None of the mutations identified so far in ABCC2 were found in these invasive populations. This truncated protein retains an intact extracellular loop 4 and thus potentially to bind to the Cry1F toxin. Further experiments are needed to determine if it can confer resistance to some of the Bt toxins.

Looking at other *Spodoptera* species, resistance to some Bt toxins was reported as early as 1994–1995 for *S. littoralis* and *S. exigua* respectively [118,119]. However, in both cases, these were laboratory selected strains. No resistance has been reported to date for *S. litura* in the Arthropod Pesticide Resistance Database although one study mentioned the development of resistance to Cry1C and Vip3A toxins in a laboratory selected strain [120]. But these studies did not involve ABCC2. In *S. exigua*, a strain was selected with a commercial *B. thuringiensis* product, Xentari. The main toxins contained in this product are Cry1a and Cry1Ca. The selected strain (Xen-R) showed a resistance factor of more than 1000 [121]. Bulk segregant analysis based on high-throughput sequencing identified the region of the genome carrying the major resistance loci, which contained three genes encoding *ABCC1*, *ABCC2* and *ABCC3* [117]. ABCC2 and ABCC3 played a major role in resistance. ABCC2 contained a mutation in the resistant strain, which resulted in the loss of the ATP binding domain II caused by the truncation of 82 terminal amino acids (Figure 10F). However, the binding of Cry1Ca to brush border membrane vesicles was not significantly different between resistant and susceptible strains. RNAi experiments confirmed the role of ABCC2 and ABCC3 in the resistance to Cry1A and Cry1Ca in *S. exigua* [117]. In additio Pinos and co-workers showed that the loss of the second nucleotide binding domain di not affect the binding of Cry1A toxins [122]. Indeed, expression of the truncated transport in Sf21 insect cells conferred sensitivity to Cry1A, as specific binding of the toxin was sti effective. These results suggest that this domain is not required for a functional toxi receptor [122]. Nevertheless, the use of techniques such as CRISPR/Cas9 validated th main role of ABCC2 as a receptor for Cry1Ac and Cry1Fa toxin in *S. exigua* [123]. The knoc out (KO) strain for SeABCC2 was more resistant than the parental strain by a factor of 47 and 240 for Cry1Ac and Cry1Fa, respectively. Using the same technique in *S. frugiperda* produce a KO for ABCC2 conferred resistance to Cry1F toxin [124].

4.1.2. Resistance through Over- or Reduced Expression of ABCs

Data on the involvement of ABC transporters in resistance to chemical insecticides i *Spodoptera* spp. are scarce, as it is in Lepidoptera in a broader sense. A few examples exis such as the involvement of *H. armigera* Pgp1 insecticide transport. HaPgp1 expressio was induced in the larval gut after exposure to abamectin [125]. When the expression c this transporter is suppressed by RNAi, larvae became more sensitive to the insecticid and mortality increased from 26% to 84% for a fixed abamectin dose of 0.4 µg/g diet [125 Another study confirmed the role of HaPgp in abamectin transport, as the use of a specif Pgp inhibitor, verapamil, increased the sensitivity of larvae towards abamectin [126].

The development of the CRISPR/Cas9 technique has paved the way to study the rol of ABCs in the transport of chemical insecticides and potentially in the development resistance. Here, we report the few examples in which an increase in LD_{50} was observe for insecticides after KO of certain ABC transporters in *Spodoptera* spp. In *S. exigua*, th ability of ABCB1 to transport 12 insecticidal molecules from 10 different chemical familie was investigated by knocking out this protein using CRISPR/Cas 9 [127]. The ABCB1 K strain showed a 2.73-fold and 3.01-fold increase in sensitivity to abamectin and emamecti benzoate, respectively, while no significant difference was observed for the other insect cides tested. In *S. frugiperda*, the KO of ABCC2 rendered the insects 7.8 and 3.1 times mor sensitive to abamectin and spinosad, respectively. Similarly, a reduction in tolerance t these two insecticides was observed for the ABCC3 KO by factors of 4.5 and 2 [128]. A other study reported that SfABCC2 KO does not induce resistance to insecticide molecule such as bifenthrin, chlorantraniliprole, spinetoram and acephate [124]. To our knowledg there are no CRISPR/Cas9-mediated ABC KO data for *S. litura* and *S. littoralis* to date.

4.1.3. Resistance and Regulation of ABCs Expression

While the mechanisms leading to the acquisition of chemotherapy resistance throug changes in ABC transporters expression of are well known in humans, those leading t pesticide resistance are still largely unexplored in insects. For example, over-expression c MDR1 (ABCB1) in human tumors has been associated with increased gene copy numbe through chromosomal region amplification, epigenetic modifications and single nucleotid polymorphisms (SNPs) [129]. In insects, the acquisition of resistance may involve eithe under- or over-expression of a given ABC transporter. Indeed, in the case of Bt toxin decreased expression of the ABC receptor can confer resistance, while conversely, a over-expression of ABCs that can transport insecticides would allow the developmen of resistance. MicroRNAs (miRNA), which are small non-coding RNA sequences (be tween 19 and 24 nucleotides in length) are known to regulate gene expression. Th miRNA miR-998-3p has target sites in the ABCC2 coding sequence (CDS) in several Lep idoptera species [130]. The involvement of these sites in the regulation of ABCC2 ha been demonstrated in *P. xylostella*. MiR-998-3p was over-expressed by a factor of two in Cry1Ac-resistant strain (GX-R) compared to a susceptible strain and ABCC2 expressio in the larval midgut was reduced by approximately 50%, demonstrating the involvemen of this miRNA in Bt resistance. A conserved target site for miR998-3p had been found i the four *Spodoptera* spp. as well as one and two non-conserved sites for *S. exigua* and th other three species respectively ([130] and the present study). MiR-998-3p could play a role in the acquisition of resistance in these species but to the best of our knowledge there has been no study has so far reported such cases and more generally the role of miRNAs in insecticide resistance remains largely unexplored.

Another regulator of *ABCC2* expression is the transcription factor Forkhead box protein A (FoxA). Its transfection into Sf9 cells (*S. frugiperda* cell model) induces the expression of ABCC2 and ABCC3 and results in increased susceptibility of the cells to Cry1Ac toxin [131]. Potential resistance could arise through under-expression of this transcription factor but has not been described at this time.

Cap 'n' collar isoform C (CncC) is a major transcription factor for controlling the expression of detoxification genes in insects [12]. CncC is constitutively overexpressed in a Chinese field population of *S. exigua* resistant to chlorpyrifos and cypermethrin [94]. Several studies on this resistant strain showed that CncC is involved in controlling the expression of detoxification genes including P450s (*CYP321A8*, *CYP321A16*, and *CYP332A1*) [11,132] and GSTs (*GSTo2*, *GSTe6*, *GSTd3*) [94]. However, none of these studies mention the role of CncC in the expression of ABC transporters. Nevertheless, this role is not excluded and an RNAseq analysis would allow a global approach to identify CncC-regulated genes in this resistant population. In *Drosophila*, a DDT-resistant strain (91R) constitutively overexpressing CncC exhibits multifactorial resistance involving some P450 and GST in addition to ABC transporters [108]. None of these studies investigated the underlying cause of CncC overexpression. In other *Spodoptera* species, CncC regulates the expression of several detoxification genes involved in indoxacarb resistance, including the ABC transporter SlituABCH-1 in *S. litura* [104]. Although the CncC gene from *S. frugiperda* has been cloned [133], it has not been shown to control of the expression of resistance-related ABC transporters and, to date, this information is also lacking in *S. littoralis*.

A number of questions remain unanswered. What is the mechanism behind the overexpression of CncC in some resistant populations? What other regulatory mechanisms could control the expression of ABC transporters involved in resistance? Do epigenetic regulations play a major role in the expression of certain ABC transporters and the acquisition of resistance, as has been observed in humans?

Supplementary Materials: The following are available online at https://www.mdpi.com/article/10.3390/insects12060544/s1, Table S1: Protein sequences of *S. littoralis* carboxyl/cholinesterases, Table S2: Spodopteran GSTes reconciled nomenclature, with new annotations combined with previous studies.

Author Contributions: All authors contribute equally to the writing. All authors have read and agreed to the published version of the manuscript.

Funding: This research received no external funding.

Acknowledgments: The authors would like to thank Abby Cuttriss (Université Côte d'Azur, Office of International Visibility). Le Goff would like to thank Dries Amezian for his very helpful comments.

Conflicts of Interest: The authors declare no conflict of interest.

References

1. Cheng, T.; Wu, J.; Wu, Y.; Chilukuri, R.V.; Huang, L.; Yamamoto, K.; Feng, L.; Li, W.; Chen, Z.; Guo, H.; et al. Genomic adaptation to polyphagy and insecticides in a major East Asian noctuid pest. *Nat. Ecol. Evol.* **2017**. [CrossRef]
2. Stokstad, E. New crop pest takes Africa at lightning speed. *Science* **2017**, *356*, 473–474. [CrossRef] [PubMed]
3. Johnson, S.J. Migration and life history strategy of the fall armyworm *Spodoptera frugiperda* in the Western hemisphere. *Insect Sci. Appl.* **1987**, *8*, 543–549. [CrossRef]
4. Westbrook, J.K. Noctuid migration in Texas within the nocturnal aeroecological boundary layer. *Integr. Comp. Biol.* **2008**, *48*, 99–106. [CrossRef] [PubMed]
5. Sparks, T.C.; Crossthwaite, A.J.; Nauen, R.; Banba, S.; Cordova, D.; Earley, F.; Ebbinghaus-Kintscher, U.; Fujioka, S.; Hirao, A.; Karmon, D.; et al. Insecticides, biologics and nematicides: Updates to IRAC's mode of action classification-a tool for resistance management. *Pestic. Biochem. Physiol.* **2020**, *167*, 104587. [CrossRef]

6. Riveron, J.M.; Yunta, C.; Ibrahim, S.S.; Djouaka, R.; Irving, H.; Menze, B.D.; Ismail, H.M.; Hemingway, J.; Ranson, H.; Albert, A. et al. A single mutation in the GSTe2 gene allows tracking of metabolically based insecticide resistance in a major malaria vector. *Genome Biol.* **2014**, *15*, R27. [CrossRef]

7. Feyereisen, R.; Dermauw, W.; van Leeuwen, T. Genotype to phenotype, the molecular and physiological dimensions of resistance in arthropods. *Pestic. Biochem. Physiol.* **2015**, *121*, 61–77. [CrossRef]

8. Schmidt, J.M.; Good, R.T.; Appleton, B.; Sherrard, J.; Raymant, G.C.; Bogwitz, M.R.; Martin, J.; Daborn, P.J.; Goddard, M.E.; Batterham, P.; et al. Copy number variation and transposable elements feature in recent, ongoing adaptation at the Cyp6g1 locus. *PLoS Genet.* **2010**, *6*, e1000998. [CrossRef] [PubMed]

9. Devonshire, A.L.; Field, L.M. Gene amplification and insecticide resistance. *Annu. Rev. Entomol.* **1991**, *36*, 1–23. [CrossRef] [PubMed]

10. Field, L.M.; Blackman, R.L.; Tyler-Smith, C.; Devonshire, A.L. Relationship between amount of esterase and gene copy number in insecticide-resistant *Myzus persicae (Sulzer)*. *Biochem. J.* **1999**, *339 Pt 3*, 737–742. [CrossRef]

11. Hu, B.; Huang, H.; Hu, S.; Ren, M.; Wei, Q.; Tian, X.; Elzaki, M.E.A.; Bass, C.; Su, J.; Palli, S.R. Changes in both trans- and cis-regulatory elements mediate insecticide resistance in a lepidopteron pest, *Spodoptera exigua*. *PLoS Genet.* **2021**, *17*, e1009403. [CrossRef] [PubMed]

12. Amezian, D.; Nauen, R.; le Goff, G. Transcriptional regulation of xenobiotic detoxification genes in insects-An overview. *Pestic. Biochem. Physiol.* **2021**, *174*, 104822. [CrossRef] [PubMed]

13. Werck-Reichhart, D.; Feyereisen, R. Cytochromes P450: A success story. *Genome Biol.* **2000**, *1*, REVIEWS3003. [CrossRef] [PubMed]

14. Mansuy, D. The great diversity of reactions catalyzed by cytochromes P450. *Comp. Biochem. Physiol. C Pharmacol. Toxicol Endocrinol.* **1998**, *121*, 5–14. [CrossRef]

15. Rewitz, K.F.; Rybczynski, R.; Warren, J.T.; Gilbert, L.I. The Halloween genes code for cytochrome P450 enzymes mediating synthesis of the insect moulting hormone. *Biochem. Soc. Trans.* **2006**, *34 Pt 6*, 1256–1260. [CrossRef]

16. Qiu, Y.; Tittiger, C.; Wicker-Thomas, C.; le Goff, G.; Young, S.; Wajnberg, E.; Fricaux, T.; Taquet, N.; Blomquist, G.J.; Feyereisen, R. An insect-specific P450 oxidative decarbonylase for cuticular hydrocarbon biosynthesis. *Proc. Natl. Acad. Sci. USA* **2012**, *109*, 14858–14863. [CrossRef]

17. Kirkness, E.F.; Haas, B.J.; Sun, W.; Braig, H.R.; Perotti, M.A.; Clark, J.M.; Lee, S.H.; Robertson, H.M.; Kennedy, R.C.; Elhaik, E. et al. Genome sequences of the human body louse and its primary endosymbiont provide insights into the permanent parasitic lifestyle. *Proc. Natl. Acad. Sci. USA* **2010**, *107*, 12168–12173. [CrossRef] [PubMed]

18. Yang, T.; Liu, N. Genome analysis of cytochrome P450s and their expression profiles in insecticide resistant mosquitoes, *Culex quinquefasciatus*. *PLoS ONE* **2011**, *6*, e29418. [CrossRef]

19. Nebert, D.W.; Adesnik, M.; Coon, M.J.; Estabrook, R.W.; Gonzalez, F.J.; Guengerich, F.P.; Gunsalus, I.C.; Johnson, E.F.; Kemper, B.; Levin, W.; et al. The P450 gene superfamily: Recommended nomenclature. *DNA* **1987**, *6*, 1–11. [CrossRef] [PubMed]

20. Nebert, D.W.; Nelson, D.R.; Coon, M.J.; Estabrook, R.W.; Feyereisen, R.; Fujii-Kuriyama, Y.; Gonzalez, F.J.; Guengerich, F.P.; Gunsalus, I.C.; Johnson, E.F.; et al. The P450 superfamily: Update on new sequences, gene mapping, and recommended nomenclature. *DNA Cell Biol.* **1991**, *10*, 1–14. [CrossRef]

21. Nelson, D.R. Metazoan cytochrome P450 evolution. *Comp. Biochem Physiol C Pharmacol Toxicol Endocrinol* **1998**, *121*, 15–22. [CrossRef]

22. Dermauw, W.; van Leeuwen, T.; Feyereisen, R. Diversity and evolution of the P450 family in arthropods. *Insect Biochem. Mol. Biol.* **2020**, *127*, 103490. [CrossRef]

23. Le Goff, G.; Boundy, S.; Daborn, P.J.; Yen, J.L.; Sofer, L.; Lind, R.; Sabourault, C.; Madi-Ravazzi, L.; ffrench-Constant, R.H. Microarray analysis of cytochrome P450 mediated insecticide resistance in *Drosophila*. *Insect Biochem. Mol. Biol.* **2003**, *33*, 701–708. [CrossRef]

24. Daborn, P.J.; Lumb, C.; Boey, A.; Wong, W.; Ffrench-Constant, R.H.; Batterham, P. Evaluating the insecticide resistance potential of eight *Drosophila melanogaster* cytochrome P450 genes by transgenic over-expression. *Insect Biochem. Mol. Biol.* **2007**, *37*, 512–519. [CrossRef] [PubMed]

25. D'Alencon, E.; Sezutsu, H.; Legeai, F.; Permal, E.; Bernard-Samain, S.; Gimenez, S.; Gagneur, C.; Cousserans, F.; Shimomura, M.; Brun-Barale, A.; et al. Extensive synteny conservation of holocentric chromosomes in Lepidoptera despite high rates of local genome rearrangements. *Proc. Natl. Acad. Sci. USA* **2010**, *107*, 7680–7685. [CrossRef]

26. Feyereisen, R. Arthropod CYPomes illustrate the tempo and mode in P450 evolution. *Biochim. Biophys. Acta* **2011**, *1814*, 19–28. [CrossRef]

27. Gimenez, S.; Abdelgaffar, H.; le Goff, G.; Hilliou, F.; Blanco, C.A.; Hanniger, S.; Bretaudeau, A.; Legeai, F.; Negre, N.; Jurat-Fuentes, J.L.; et al. Adaptation by copy number variation increases insecticide resistance in the fall armyworm. *Commun. Biol.* **2020**, *3*, 664. [CrossRef] [PubMed]

28. Gouin, A.; Bretaudeau, A.; Nam, K.; Gimenez, S.; Aury, J.M.; Duvic, B.; Hilliou, F.; Durand, N.; Montagne, N.; Darboux, I.; et al. Two genomes of highly polyphagous lepidopteran pests (*Spodoptera frugiperda*, Noctuidae) with different host-plant ranges. *Sci. Rep.* **2017**, *7*, 11816. [CrossRef] [PubMed]

29. Kergoat, G.J.; Goldstein, P.Z.; le Ru, B.; Meagher, R.L., Jr.; Zilli, A.; Mitchell, A.; Clamens, A.L.; Gimenez, S.; Barbut, J.; Negre, N.; et al. A novel reference dated phylogeny for the genus *Spodoptera* Guenee (Lepidoptera: Noctuidae: Noctuinae): New insights into the evolution of a pest-rich genus. *Mol. Phylogenet. Evol.* **2021**, 107161. [CrossRef]

0.   Klai, K.; Chenais, B.; Zidi, M.; Djebbi, S.; Caruso, A.; Denis, F.; Confais, J.; Badawi, M.; Casse, N.; Khemakhem, M.M. Screening of *Helicoverpa armigera* mobilome revealed transposable element insertions in insecticide resistance genes. *Insects* **2020**, *11*, 879. [CrossRef] [PubMed]

1.   Yu, L.; Tang, W.; He, W.; Ma, X.; Vasseur, L.; Baxter, S.W.; Yang, G.; Huang, S.; Song, F.; You, M. Characterization and expression of the cytochrome P450 gene family in diamondback moth, *Plutella xylostella* (L.). *Sci. Rep.* **2015**, *5*, 8952. [CrossRef] [PubMed]

2.   Carareto, C.M.; Hernandez, E.H.; Vieira, C. Genomic regions harboring insecticide resistance-associated Cyp genes are enriched by transposable element fragments carrying putative transcription factor binding sites in two sibling *Drosophila* species. *Gene* **2014**, *537*, 93–99. [CrossRef] [PubMed]

3.   Wilding, C.S.; Smith, I.; Lynd, A.; Yawson, A.E.; Weetman, D.; Paine, M.J.; Donnelly, M.J. A cis-regulatory sequence driving metabolic insecticide resistance in mosquitoes: Functional characterisation and signatures of selection. *Insect Biochem. Mol. Biol.* **2012**, *42*, 699–707. [CrossRef] [PubMed]

4.   Cheng, D.; Guo, Z.; Riegler, M.; Xi, Z.; Liang, G.; Xu, Y. Gut symbiont enhances insecticide resistance in a significant pest, the oriental fruit fly *Bactrocera dorsalis* (Hendel). *Microbiome* **2017**, *5*, 13. [CrossRef]

5.   Giraudo, M.; Hilliou, F.; Fricaux, T.; Audant, P.; Feyereisen, R.; le Goff, G. Cytochrome P450s from the fall armyworm (*Spodoptera frugiperda*): Responses to plant allelochemicals and pesticides. *Insect Mol. Biol.* **2015**, *24*, 115–128. [CrossRef] [PubMed]

6.   Shi, Y.; Wang, H.; Liu, Z.; Wu, S.; Yang, Y.; Feyereisen, R.; Heckel, D.G.; Wu, Y. Phylogenetic and functional characterization of ten P450 genes from the CYP6AE subfamily of *Helicoverpa armigera* involved in xenobiotic metabolism. *Insect Biochem. Mol. Biol.* **2018**, *93*, 79–91. [CrossRef] [PubMed]

7.   Rose, R.L.; Goh, D.; Thompson, D.M.; Verma, K.D.; Heckel, D.G.; Gahan, L.J.; Roe, R.M.; Hodgson, E. Cytochrome P450 (CYP)9A1 in *Heliothis virescens*: The first member of a new CYP family. *Insect Biochem. Mol. Biol.* **1997**, *27*, 605–615. [CrossRef]

8.   Yang, Y.; Chen, S.; Wu, S.; Yue, L.; Wu, Y. Constitutive overexpression of multiple cytochrome P450 genes associated with pyrethroid resistance in *Helicoverpa armigera*. *J. Econ. Entomol.* **2006**, *99*, 1784–1789. [CrossRef]

9.   Yang, Y.; Yue, L.; Chen, S.; Wu, Y. Functional expression of Helicoverpa armigera CYP9A12 and CYP9A14 in Saccharomyces cerevisiae. *Pestic. Biochem. Physiol.* **2008**, *92*, 101–105. [CrossRef]

0.   Boaventura, D.; Buer, B.; Hamaekers, N.; Maiwald, F.; Nauen, R. Toxicological and molecular profiling of insecticide resistance in a Brazilian strain of fall armyworm resistant to Bt Cry1 proteins. *Pest. Manag. Sci.* **2020**. [CrossRef]

1.   Do Nascimento, A.R.; Fresia, P.; Consoli, F.L.; Omoto, C. Comparative transcriptome analysis of lufenuron-resistant and susceptible strains of *Spodoptera frugiperda* (Lepidoptera: Noctuidae). *BMC Genom.* **2015**, *16*, 985. [CrossRef] [PubMed]

2.   Hafeez, M.; Liu, S.; Yousaf, H.K.; Jan, S.; Wang, R.L.; Fernandez-Grandon, G.M.; Li, X.; Gulzar, A.; Ali, B.; Rehman, M.; et al. RNA interference-mediated knockdown of a cytochrome P450 gene enhanced the toxicity of alpha-cypermethrin in xanthotoxin-fed larvae of *Spodoptera exigua* (Hubner). *Pestic. Biochem. Physiol.* **2020**, *162*, 6–14. [CrossRef]

3.   Hafeez, M.; Liu, S.; Jan, S.; Shi, L.; Fernandez-Grandon, G.M.; Gulzar, A.; Ali, B.; Rehman, M.; Wang, M. Knock-down of gossypol-inducing cytochrome P450 genes reduced deltamethrin sensitivity in *Spodoptera exigua* (Hubner). *Int. J. Mol. Sci.* **2019**, *20*, 2248. [CrossRef]

4.   Wang, R.L.; Liu, S.W.; Baerson, S.R.; Qin, Z.; Ma, Z.H.; Su, Y.J.; Zhang, J.E. Identification and functional analysis of a novel cytochrome P450 gene CYP9A105 associated with pyrethroid detoxification in *Spodoptera exigua* Hubner. *Int. J. Mol. Sci.* **2018**, *19*, 737. [CrossRef]

5.   Xu, L.; Mei, Y.; Liu, R.; Chen, X.; Li, D.; Wang, C. Transcriptome analysis of *Spodoptera litura* reveals the molecular mechanism to pyrethroids resistance. *Pestic. Biochem. Physiol.* **2020**, *169*, 104649. [CrossRef] [PubMed]

6.   Wang, R.L.; Staehelin, C.; Xia, Q.Q.; Su, Y.J.; Zeng, R.S. Identification and characterization of CYP9A40 from the tobacco cutworm moth (*Spodoptera litura*), a cytochrome P450 gene induced by plant allelochemicals and insecticides. *Int. J. Mol. Sci.* **2015**, *16*, 22606–22620. [CrossRef] [PubMed]

7.   Oakeshott, J.; Claudianos, C.; Campbell, P.M.; Newcomb, R.; Russell, R.J. Biochemical genetics and genomics of insect esterases. In *Comprehensive Molecular Insect Science-Pharmacology*; Gilbert, L.I., Iatrou, K., Gill, S.S., Eds.; Elsevier: Amsterdam, The Netherlands, 2005; pp. 309–381.

8.   Li, X.; Schuler, M.A.; Berenbaum, M.R. Molecular mechanisms of metabolic resistance to synthetic and natural xenobiotics. *Annu. Rev. Entomol.* **2007**, *52*, 231–253. [CrossRef]

9.   Oakeshott, J.G.; Devonshire, A.L.; Claudianos, C.; Sutherland, T.D.; Horne, I.; Campbell, P.M.; Ollis, D.L.; Russell, R.J. Comparing the organophosphorus and carbamate insecticide resistance mutations in cholin- and carboxyl-esterases. *Chem. Biol. Interact.* **2005**, *157–158*, 269–275. [CrossRef]

50.   Oakeshott, J.G.; Johnson, R.M.; Berenbaum, M.R.; Ranson, H.; Cristino, A.S.; Claudianos, C. Metabolic enzymes associated with xenobiotic and chemosensory responses in *Nasonia vitripennis*. *Insect Mol. Biol.* **2010**, *19* (Suppl. 1), 147–163. [CrossRef]

51.   Pearce, S.L.; Clarke, D.F.; East, P.D.; Elfekih, S.; Gordon, K.H.J.; Jermiin, L.S.; McGaughran, A.; Oakeshott, J.G.; Papanikolaou, A.; Perera, O.P.; et al. Genomic innovations, transcriptional plasticity and gene loss underlying the evolution and divergence of two highly polyphagous and invasive *Helicoverpa* pest species. *BMC Biol.* **2017**, *15*, 63. [CrossRef]

52.   Durand, N.; Carot-Sans, G.; Chertemps, T.; Montagne, N.; Jacquin-Joly, E.; Debernard, S.; Maibeche-Coisne, M. A diversity of putative carboxylesterases are expressed in the antennae of the noctuid moth *Spodoptera littoralis*. *Insect Mol. Biol.* **2010**, *19*, 87–97. [CrossRef] [PubMed]

53. Durand, N.; Chertemps, T.; Maibeche-Coisne, M. Antennal carboxylesterases in a moth, structural and functional diversi* *Commun. Integr. Biol.* **2012**, *5*, 284–286. [CrossRef] [PubMed]

54. Merlin, C.; Rosell, G.; Carot-Sans, G.; Francois, M.C.; Bozzolan, F.; Pelletier, J.; Jacquin-Joly, E.; Guerrero, A.; Maibeche-Coisne, I Antennal esterase cDNAs from two pest moths, *Spodoptera littoralis* and *Sesamia nonagrioides*, potentially involved in odoura degradation. *Insect Mol. Biol.* **2007**, *16*, 73–81. [CrossRef]

55. Walker, W.B., 3rd; Roy, A.; Anderson, P.; Schlyter, F.; Hansson, B.S.; Larsson, M.C. Transcriptome Analysis of gene famili involved in chemosensory function in *Spodoptera littoralis* (Lepidoptera: Noctuidae). *BMC Genom.* **2019**, *20*, 428. [CrossRe [PubMed]

56. Farnsworth, C.A.; Teese, M.G.; Yuan, G.; Li, Y.; Scott, C.; Zhang, X.; Wu, Y.; Russell, R.J.; Oakeshott, J.G. Esterase-based metabol resistance to insecticides in heliothine and spodopteran pests. *J. Pestic. Sci.* **2010**, *35*, 275–289. [CrossRef]

57. El-Guindy, M.A.; Saleh, W.S.; El-Refai, A.R.A.; Abou-Donia, S.A. The role of heamolymph esterases as protectants again intoxication by fenitrothion in the cotton leafworm *Spodoptera littoralis* (Boisd.). *Bull. Entomol. Soc. Egypt. Econ. Ser.* **198** *14*, 177–197.

58. Cho, J.R.; Kim, Y.J.; Kim, J.J.; Kim, H.S.; Yoo, J.K.; Lee, J.O. Electrophoretic Pattern of Larval Esterases in field and laborator selected strains of the tobacco cutworm, *Spodoptera litura* (Fabricius). *J. Asia-Pac. Entomol.* **1999**, *2*, 39–44. [CrossRef]

59. Kim, Y.G.; Lee, J.I.; Kang, S.Y.; Han, S.C. Variation in insecticide susceptibilities of the beet armyworm, *Spodoptera exigua* (Hubne Esterase and acetylcholinesterase activities. *Korean J. Appl. Entomol.* **1997**, *36*, 172–178.

60. Yu, S.J.; Nguyen, S.N.; Abo-Elghar, G.E. Biochemical characteristics of insecticide resistance in the fall armyworm, *Spodopte frugiperda* (J.E. Smith). *Pestic. Biochem. Physiol.* **2003**, *77*, 1–11. [CrossRef]

61. Han, Y.; Wu, S.; Li, Y.; Liu, J.; Campbell, P.M.; Farnsworth, C.A.; Scott, C.; Russell, R.J.; Oakeshott, J.G.; Wu, Y. Proteomic an molecular analyses of esterases associated with monocrotophos resistance in *Helicoverpa armigera*. *Pestic. Biochem. Physiol.* **201** *104*, 243–251. [CrossRef]

62. Li, Y.; Farnsworth, C.A.; Coppin, C.W.; Teese, M.G.; Liu, J.W.; Scott, C.; Zhang, X.; Russell, R.J.; Oakeshott, J.G. Organophospha and pyrethroid hydrolase activities of mutant Esterases from the cotton bollworm *Helicoverpa armigera*. *PLoS ONE* **2013**, *8*, e7768 [CrossRef]

63. Teese, M.G.; Farnsworth, C.A.; Li, Y.; Coppin, C.W.; Devonshire, A.L.; Scott, C.; East, P.; Russell, R.J.; Oakeshott, J.G. Heterologou expression and biochemical characterisation of fourteen esterases from *Helicoverpa armigera*. *PLoS ONE* **2013**, *8*, e65951. [CrossRe

64. Campbell, P.M.; Trott, J.F.; Claudianos, C.; Smyth, K.A.; Russell, R.J.; Oakeshott, J.G. Biochemistry of esterases associated wit organophosphate resistance in *Lucilia cuprina* with comparisons to putative orthologues in other Diptera. *Biochem. Genet.* **199** *35*, 17–40. [CrossRef]

65. Claudianos, C.; Russell, R.J.; Oakeshott, J.G. The same amino acid substitution in orthologous esterases confers organophospha resistance on the house fly and a blowfly. *Insect Biochem. Mol. Biol.* **1999**, *29*, 675–686. [CrossRef]

66. Newcomb, R.D.; Campbell, P.M.; Ollis, D.L.; Cheah, E.; Russell, R.J.; Oakeshott, J.G. A single amino acid substitution converts carboxylesterase to an organophosphorus hydrolase and confers insecticide resistance on a blowfly. *Proc. Natl. Acad. Sci. US.* **1997**, *94*, 7464–7468. [CrossRef]

67. Cui, F.; Lin, Z.; Wang, H.; Liu, S.; Chang, H.; Reeck, G.; Qiao, C.; Raymond, M.; Kang, L. Two single mutations commonly caus qualitative change of nonspecific carboxylesterases in insects. *Insect Biochem. Mol. Biol.* **2011**, *41*, 1–8. [CrossRef]

68. Zhu, Y.; Dowdy, A.K.; Baker, J.E. Detection of single-base substitution in an esterase gene and its linkage to malathion resistanc in the parasitoid *Anisopteromalus calandrae* (Hymenoptera: Pteromalidae). *Pestic. Sci.* **1999**, *55*, 398–404. [CrossRef]

69. Fournier, D. Mutations of acetylcholinesterase which confer insecticide resistance in insect populations. *Chem. Biol. Interact.* **200** *157–158*, 257–261. [CrossRef] [PubMed]

70. Andrews, M.C.; Callaghan, A.; Field, L.M.; Williamson, M.S.; Moores, G.D. Identification of mutations conferring insecticid insensitive AChE in the cotton-melon aphid, *Aphis gossypii* Glover. *Insect Mol. Biol.* **2004**, *13*, 555–561. [CrossRef]

71. Hsu, J.C.; Haymer, D.S.; Wu, W.J.; Feng, H.T. Mutations in the acetylcholinesterase gene of *Bactrocera dorsalis* associated wit resistance to organophosphorus insecticides. *Insect Biochem. Mol. Biol.* **2006**, *36*, 396–402. [CrossRef] [PubMed]

72. Jiang, X.; Qu, M.; Denholm, I.; Fang, J.; Jiang, W.; Han, Z. Mutation in acetylcholinesterase1 associated with triazophos resistanc in rice stem borer, *Chilo suppressalis* (Lepidoptera: Pyralidae). *Biochem. Biophys. Res. Commun.* **2009**, *378*, 269–272. [CrossRe [PubMed]

73. Lee, D.W.; Choi, J.Y.; Kim, W.T.; Je, Y.H.; Song, J.T.; Chung, B.K.; Boo, K.S.; Koh, Y.H. Mutations of acetylcholinesterase1 contribut to prothiofos-resistance in *Plutella xylostella* (L.). *Biochem. Biophys. Res. Commun.* **2007**, *353*, 591–597. [CrossRef] [PubMed]

74. Kim, Y.; Cho, J.R.; Lee, J.I.; Kang, S.Y.; Han, S.C.; Hong, K.J.; Kim, H.S.; Yoo, J.K.; Lee, J.O. Insecticide resistance in the tobacc cutworm, *Spodoptera litura* (Fabricius) (Lepidoptera: Noctuidae). *J. Asia-Pac. Entomol.* **1998**, *1*, 115–122.

75. Kranthi, K.R.; Jadhav, D.R.; Kranthi, S.; Wanjari, R.R.; Ali, S.S.; Russell, D.A. Insecticide resistance in five major insect pests c cotton in India. *Crop. Prot.* **2002**, *21*, 449–460. [CrossRef]

76. Muthusamy, Shivakumar, Karthi, and Ramkumar, Pesticide detoxifying mechanism in field population of *Spodoptera litur* (Lepidoptera: Noctuidae) from South India *Egypt. Acad. J. Biolog. Sci.* **2011**, *3*, 51–57.

77. Byrne, F.J.; Toscano, N.C. An insensitive acetylcholinesterase confers resistance to methomyl in the beet armyworm *Spodopter exigua* (Lepidoptera: Noctuidae). *J. Econ. Entomol.* **2001**, *94*, 524–528. [CrossRef]

78. Carvalho, R.A.; Omoto, C.; Field, L.M.; Williamson, M.S.; Bass, C. Investigating the molecular mechanisms of organophosphate and pyrethroid resistance in the fall armyworm *Spodoptera frugiperda*. *PLoS ONE* **2013**, *8*, e62268. [CrossRef]

79. Boaventura, D.; Martin, M.; Pozzebon, A.; Mota-Sanchez, D.; Nauen, R. Monitoring of target-site mutations conferring insecticide resistance in *Spodoptera frugiperda*. *Insects* **2020**, *11*, 545. [CrossRef]

80. Guan, F.; Zhang, J.; Shen, H.; Wang, X.; Padovan, A.; Walsh, T.K.; Tay, W.T.; Gordon, K.H.J.; James, W.; Czepak, C.; et al. Whole-genome sequencing to detect mutations associated with resistance to insecticides and Bt proteins in *Spodoptera frugiperda*. *Insect Sci.* **2020**. [CrossRef]

81. Jakoby, W.B.; Ziegler, D.M. The enzymes of detoxication. *J. Biol. Chem.* **1990**, *265*, 20715–20718. [CrossRef]

82. Enayati, A.A.; Ranson, H.; Hemingway, J. Insect glutathione transferases and insecticide resistance. *Insect Mol. Biol.* **2005**, *14*, 3–8. [CrossRef] [PubMed]

83. Sheehan, D.; Meade, G.; Foley, V.M.; Dowd, C.A. Structure, function and evolution of glutathione transferases: Implications for classification of non-mammalian members of an ancient enzyme superfamily. *Biochem. J.* **2001**, *360 Pt 1*, 1–16. [CrossRef]

84. Hu, B.; Hu, S.; Huang, H.; Wei, Q.; Ren, M.; Huang, S.; Tian, X.; Su, J. Insecticides induce the co-expression of glutathione S-transferases through ROS/CncC pathway in *Spodoptera exigua*. *Pestic. Biochem. Physiol.* **2019**, *155*, 58–71. [CrossRef]

85. Friedman, R. Genomic organization of the glutathione S-transferase family in insects. *Mol. Phylogenet. Evol.* **2011**, *61*, 924–932. [CrossRef] [PubMed]

86. Kostaropoulos, I.; Papadopoulos, A.I.; Metaxakis, A.; Boukouvala, E.; Papadopoulou-Mourkidou, E. Glutathione S-transferase in the defence against pyrethroids in insects. *Insect Biochem. Mol. Biol.* **2001**, *31*, 313–319. [CrossRef]

87. Alias, Z.; Clark, A.G. Studies on the glutathione S-transferase proteome of adult *Drosophila melanogaster*: Responsiveness to chemical challenge. *Proteomics* **2007**, *7*, 3618–3628. [CrossRef] [PubMed]

88. Lumjuan, N.; Rajatileka, S.; Changsom, D.; Wicheer, J.; Leelapat, P.; Prapanthadara, L.A.; Somboon, P.; Lycett, G.; Ranson, H. The role of the *Aedes aegypti* Epsilon glutathione transferases in conferring resistance to DDT and pyrethroid insecticides. *Insect Biochem. Mol. Biol.* **2011**, *41*, 203–209. [CrossRef]

89. Hardy, N.B.; Peterson, D.A.; Ross, L.; Rosenheim, J.A. Does a plant-eating insect's diet govern the evolution of insecticide resistance? Comparative tests of the pre-adaptation hypothesis. *Evol. Appl.* **2018**, *11*, 739–747. [CrossRef]

90. Pavlidi, N.; Vontas, J.; van Leeuwen, T. The role of glutathione S-transferases (GSTs) in insecticide resistance in crop pests and disease vectors. *Curr. Opin. Insect Sci.* **2018**, *27*, 97–102. [CrossRef] [PubMed]

91. Vontas, J.G.; Small, G.J.; Hemingway, J. Glutathione S-transferases as antioxidant defence agents confer pyrethroid resistance in *Nilaparvata lugens*. *Biochem. J.* **2001**, *357 Pt 1*, 65–72. [CrossRef]

92. Deng, H.; Huang, Y.; Feng, Q.; Zheng, S. Two epsilon glutathione S-transferase cDNAs from the common cutworm, *Spodoptera litura*: Characterization and developmental and induced expression by insecticides. *J. Insect Physiol.* **2009**, *55*, 1174–1183. [CrossRef] [PubMed]

93. Wan, H.; Zhan, S.; Xia, X.; Xu, P.; You, H.; Jin, B.R.; Li, J. Identification and functional characterization of an epsilon glutathione S-transferase from the beet armyworm (*Spodoptera exigua*). *Pestic. Biochem. Physiol.* **2016**, *132*, 81–88. [CrossRef] [PubMed]

94. Hu, B.; Huang, H.; Wei, Q.; Ren, M.; Mburu, D.K.; Tian, X.; Su, J. Transcription factors CncC/Maf and AhR/ARNT coordinately regulate the expression of multiple GSTs conferring resistance to chlorpyrifos and cypermethrin in *Spodoptera exigua*. *Pest. Manag. Sci.* **2019**, *75*, 2009–2019. [CrossRef]

95. Xu, Z.B.; Zou, X.P.; Zhang, N.; Feng, Q.L.; Zheng, S.C. Detoxification of insecticides, allechemicals and heavy metals by glutathione S-transferase SlGSTE1 in the gut of *Spodoptera litura*. *Insect Sci.* **2015**, *22*, 503–511. [CrossRef] [PubMed]

96. Zou, X.; Xu, Z.; Zou, H.; Liu, J.; Chen, S.; Feng, Q.; Zheng, S. Glutathione S-transferase SlGSTE1 in *Spodoptera litura* may be associated with feeding adaptation of host plants. *Insect Biochem Mol. Biol.* **2016**, *70*, 32–43. [CrossRef]

97. Hirowatari, A.; Chen, Z.; Mita, K.; Yamamoto, K. Enzymatic characterization of two epsilon-class glutathione S-transferases of *Spodoptera litura*. *Arch. Insect Biochem. Physiol.* **2018**, *97*. [CrossRef]

98. Huang, Y.; Xu, Z.; Lin, X.; Feng, Q.; Zheng, S. Structure and expression of glutathione S-transferase genes from the midgut of the common cutworm, *Spodoptera litura* (Noctuidae) and their response to xenobiotic compounds and bacteria. *J. Insect Physiol.* **2011**, *57*, 1033–1044. [CrossRef]

99. Zhang, N.; Liu, J.; Chen, S.N.; Huang, L.H.; Feng, Q.L.; Zheng, S.C. Expression profiles of glutathione S-transferase superfamily in *Spodoptera litura* tolerated to sublethal doses of chlorpyrifos. *Insect Sci.* **2016**, *23*, 675–687. [CrossRef]

100. Lalouette, L.; Pottier, M.A.; Wycke, M.A.; Boitard, C.; Bozzolan, F.; Maria, A.; Demondion, E.; Chertemps, T.; Lucas, P.; Renault, D.; et al. Unexpected effects of sublethal doses of insecticide on the peripheral olfactory response and sexual behavioral in a pest insect. *Environ. Sci. Pollut. Res.* **2016**, *23*, 3073–3085. [CrossRef]

101. Chen, X.; Zhang, Y.L. Identification and characterisation of multiple glutathione S-transferase genes from the diamondback moth, *Plutella xylostella*. *Pest. Manag. Sci.* **2015**, *71*, 592–600. [CrossRef]

102. Huang, H.S.; Hu, N.T.; Yao, Y.E.; Wu, C.Y.; Chiang, S.W.; Sun, C.N. Molecular cloning and heterologous expression of a glutathione S-transferase involved in insecticide resistance from the diamondback moth, *Plutella xylostella*. *Insect Biochem. Mol. Biol.* **1998**, *28*, 651–658. [CrossRef]

103. Shi, L.; Shi, Y.; Zhang, Y.; Liao, X. A systemic study of indoxacarb resistance in *Spodoptera litura* revealed complex expression profiles and regulatory mechanism. *Sci. Rep.* **2019**, *9*, 14997. [CrossRef]

104. Shi, L.; Shi, Y.; Liu, M.F.; Zhang, Y.; Liao, X.L. Transcription factor CncC potentially regulates the expression of multiple detoxification genes that mediate indoxacarb resistance in *Spodoptera litura*. *Insect Sci.* **2020**. [CrossRef]

105. Li, Z.; Cai, T.; Qin, Y.; Zhang, Y.; Jin, R.; Mao, K.; Liao, X.; Wan, H.; Li, J. Transcriptional Response of ATP-Binding Cassette (ABC) Transporters to Insecticide in the Brown Planthopper, *Nilaparvata lugens* (Stal). *Insects* **2020**, *11*, 280. [CrossRef]

106. Qi, W.; Ma, X.; He, W.; Chen, W.; Zou, M.; Gurr, G.M.; Vasseur, L.; You, M. Characterization and expression profiling of ATP-binding cassette transporter genes in the diamondback moth, *Plutella xylostella* (L.). *BMC Genom.* **2016**, *17*, 760. [CrossRef]

107. Briz, O.; Perez-Silva, L.; Al-Abdulla, R.; Abete, L.; Reviejo, M.; Romero, M.R.; Marin, J.J.G. What "The Cancer Genome Atlas" database tells us about the role of ATP-binding cassette (ABC) proteins in chemoresistance to anticancer drugs. *Expert Opin Drug Metab. Toxicol.* **2019**, *15*, 577–593. [CrossRef] [PubMed]

108. Gellatly, K.J.; Yoon, K.S.; Doherty, J.J.; Sun, W.; Pittendrigh, B.R.; Clark, J.M. RNAi validation of resistance genes and their interactions in the highly DDT-resistant 91-R strain of *Drosophila melanogaster*. *Pestic. Biochem. Physiol.* **2015**, *121*, 107–115. [CrossRef] [PubMed]

109. Jurat-Fuentes, J.L.; Heckel, D.G.; Ferre, J. Mechanisms of Resistance to Insecticidal Proteins from *Bacillus thuringiensis*. *Annu. Rev Entomol.* **2021**, *66*, 121–140. [CrossRef] [PubMed]

110. Gahan, L.J.; Pauchet, Y.; Vogel, H.; Heckel, D.G. An ABC transporter mutation is correlated with insect resistance to *Bacillus thuringiensis* Cry1Ac toxin. *PLoS Genet.* **2010**, *6*, e1001248. [CrossRef]

111. Liu, L.; Chen, Z.; Yang, Y.; Xiao, Y.; Liu, C.; Ma, Y.; Soberon, M.; Bravo, A.; Yang, Y.; Liu, K. A single amino acid polymorphism in ABCC2 loop 1 is responsible for differential toxicity of *Bacillus thuringiensis* Cry1Ac toxin in different *Spodoptera* (Noctuidae) species. *Insect Biochem. Mol. Biol.* **2018**, *100*, 59–65. [CrossRef] [PubMed]

112. Endo, H.; Tanaka, S.; Adegawa, S.; Ichino, F.; Tabunoki, H.; Kikuta, S.; Sato, R. Extracellular loop structures in silkworm ABCC transporters determine their specificities for *Bacillus thuringiensis* Cry toxins. *J. Biol. Chem.* **2018**, *293*, 8569–8577. [CrossRef] [PubMed]

113. De Maagd, R.A.; Bravo, A.; Crickmore, N. How *Bacillus thuringiensis* has evolved specific toxins to colonize the insect world *Trends Genet.* **2001**, *17*, 193–199. [CrossRef]

114. Banerjee, R.; Hasler, J.; Meagher, R.; Nagoshi, R.; Hietala, L.; Huang, F.; Narva, K.; Jurat-Fuentes, J.L. Mechanism and DNA-based detection of field-evolved resistance to transgenic Bt corn in fall armyworm (*Spodoptera frugiperda*). *Sci. Rep.* **2017**, *7*, 10877 [CrossRef] [PubMed]

115. Flagel, L.; Lee, Y.W.; Wanjugi, H.; Swarup, S.; Brown, A.; Wang, J.; Kraft, E.; Greenplate, J.; Simmons, J.; Adams, N.; et al Mutational disruption of the ABCC2 gene in fall armyworm, *Spodoptera frugiperda*, confers resistance to the Cry1Fa and Cry1A.105 insecticidal proteins. *Sci. Rep.* **2018**, *8*, 7255. [CrossRef]

116. Boaventura, D.; Ulrich, J.; Lueke, B.; Bolzan, A.; Okuma, D.; Gutbrod, O.; Geibel, S.; Zeng, Q.; Dourado, P.M.; Martinelli, S.; et al Molecular characterization of Cry1F resistance in fall armyworm, *Spodoptera frugiperda* from Brazil. *Insect Biochem. Mol. Biol.* **2020**, *116*, 103280. [CrossRef]

117. Park, Y.; Gonzalez-Martinez, R.M.; Navarro-Cerrillo, G.; Chakroun, M.; Kim, Y.; Ziarsolo, P.; Blanca, J.; Canizares, J.; Ferre, J.; Herrero, S. ABCC transporters mediate insect resistance to multiple Bt toxins revealed by bulk segregant analysis. *BMC Biol* **2014**, *12*, 46. [CrossRef]

118. Moar, W.J.; Pusztai-Carey, M.; van Faassen, H.; Bosch, D.; Frutos, R.; Rang, C.; Luo, K.; Adang, M.J. Development of *Bacillus thuringiensis* CryIC resistance by *Spodoptera exigua* (Hubner) (Lepidoptera: Noctuidae). *Appl. Environ. Microbiol.* **1995**, *61*, 2086–2092 [CrossRef]

119. Müller-Cohn, J.; Chaufaux, J.; Buisson, C.; Gilois, N.; Sanchis, V.; Lereclus, D. *Spodoptera littoralis* (Lepidoptera: Noctuidae) resistance to CryIC and cross-resistance to other *Bacillus thuringiensis* crystal toxins. *J. Econ. Entomol.* **1996**, *89*, 791–797. [CrossRef]

120. Barkhade, U.P.; Thakare, A.S. Protease mediated resistance mechanism to Cry1C and Vip3A in *Spodoptera litura*. *Egypt. Acad. J. Biol. Sci.* **2010**, *6*, 43–50. [CrossRef]

121. Hernandez-Martinez, P.; Navarro-Cerrillo, G.; Caccia, S.; de Maagd, R.A.; Moar, W.J.; Ferre, J.; Escriche, B.; Herrero, S. Constitutive activation of the midgut response to *Bacillus thuringiensis* in Bt-resistant *Spodoptera exigua*. *PLoS ONE* **2010**, *5*. [CrossRef]

122. Pinos, D.; Martinez-Solis, M.; Herrero, S.; Ferre, J.; Hernandez-Martinez, P. The *Spodoptera exigua* ABCC2 acts as a Cry1A receptor independently of its nucleotide binding domain II. *Toxins* **2019**, *11*, 172. [CrossRef] [PubMed]

123. Huang, J.; Xu, Y.; Zuo, Y.; Yang, Y.; Tabashnik, B.E.; Wu, Y. Evaluation of five candidate receptors for three Bt toxins in the beet armyworm using CRISPR-mediated gene knockouts. *Insect Biochem. Mol. Biol.* **2020**, *121*, 103361. [CrossRef]

124. Abdelgaffar, H.; Perera, O.P.; Jurat-Fuentes, J.L. ABC transporter mutations in Cry1F-resistant fall armyworm (*Spodoptera frugiperda*) do not result in altered susceptibility to selected small molecule pesticides. *Pest. Manag. Sci.* **2021**, *77*, 949–955. [CrossRef]

125. Xiang, M.; Zhang, L.; Lu, Y.; Tang, Q.; Liang, P.; Shi, X.; Song, D.; Gao, X. A P-glycoprotein gene serves as a component of the protective mechanisms against 2-tridecanone and abamectin in *Helicoverpa armigera*. *Gene* **2017**, *627*, 63–71. [CrossRef]

126. Jin, M.; Liao, C.; Chakrabarty, S.; Zheng, W.; Wu, K.; Xiao, Y. Transcriptional response of ATP-binding cassette (ABC) transporters to insecticides in the cotton bollworm, *Helicoverpa armigera*. *Pestic. Biochem. Physiol.* **2019**, *154*, 46–59. [CrossRef]

127. Zuo, Y.Y.; Huang, J.L.; Wang, J.; Feng, Y.; Han, T.T.; Wu, Y.D.; Yang, Y.H. Knockout of a P-glycoprotein gene increases susceptibility to abamectin and emamectin benzoate in *Spodoptera exigua*. *Insect Mol. Biol.* **2017**. [CrossRef]

28. Jin, M.; Yang, Y.; Shan, Y.; Chakrabarty, S.; Cheng, Y.; Soberon, M.; Bravo, A.; Liu, K.; Wu, K.; Xiao, Y. Two ABC transporters are differentially involved in the toxicity of two *Bacillus thuringiensis* Cry1 toxins to the invasive crop-pest *Spodoptera frugiperda* (J. E. Smith). *Pest. Manag. Sci.* **2021**, *77*, 1492–1501. [CrossRef]

29. Genovese, I.; Ilari, A.; Assaraf, Y.G.; Fazi, F.; Colotti, G. Not only P-glycoprotein: Amplification of the ABCB1-containing chromosome region 7q21 confers multidrug resistance upon cancer cells by coordinated overexpression of an assortment of resistance-related proteins. *Drug Resist. Updat.* **2017**, *32*, 23–46. [CrossRef]

30. Zhu, B.; Sun, X.; Nie, X.; Liang, P.; Gao, X. MicroRNA-998-3p contributes to Cry1Ac-resistance by targeting ABCC2 in lepidopteran insects. *Insect Biochem. Mol. Biol.* **2020**, *117*, 103283. [CrossRef] [PubMed]

31. Li, J.; Ma, Y.; Yuan, W.; Xiao, Y.; Liu, C.; Wang, J.; Peng, J.; Peng, R.; Soberon, M.; Bravo, A.; et al. FOXA transcriptional factor modulates insect susceptibility to *Bacillus thuringiensis* Cry1Ac toxin by regulating the expression of toxin-receptor ABCC2 and ABCC3 genes. *Insect Biochem. Mol. Biol.* **2017**, *88*, 1–11. [CrossRef] [PubMed]

32. Bo, H.; Miaomiao, R.; Jianfeng, F.; Sufang, H.; Xia, W.; Elzaki, M.E.A.; Chris, B.; Palli, S.R.; Jianya, S. Xenobiotic transcription factors CncC and maf regulate expression of CYP321A16 and CYP332A1 that mediate chlorpyrifos resistance in *Spodoptera exigua*. *J. Hazard. Mater.* **2020**, *398*, 122971. [CrossRef] [PubMed]

33. Pan, Y.; Zeng, X.; Wen, S.; Liu, X.; Shang, Q. Characterization of the Cap 'n' Collar Isoform C gene in *Spodoptera frugiperda* and its Association with Superoxide Dismutase. *Insects* **2020**, *11*, 221. [CrossRef] [PubMed]

MDPI
St. Alban-Anlage 66
4052 Basel
Switzerland
Tel. +41 61 683 77 34
Fax +41 61 302 89 18
www.mdpi.com

Insects Editorial Office
E-mail: insects@mdpi.com
www.mdpi.com/journal/insects